A Framework for Enhancing Airlift Planning and Execution Capabilities Within the Joint Expeditionary Movement System

Robert S. Tripp, Kristin F. Lynch, Charles Robert Roll, Jr.,
John G. Drew, Patrick Mills

Prepared for the United States Air Force

Approved for public release, distribution unlimited

RAND PROJECT AIR FORCE

The research described in this report was sponsored by the United States Air Force under Contract F49642-01-C-0003. Further information may be obtained from the Strategic Planning Division, Directorate of Plans, Hq USAF.

Library of Congress Cataloging-in-Publication Data

A framework for enhancing airlift planning and execution capabilities within the joint expeditionary movement system / Robert S. Tripp ... [et al.].

p. cm.

"MG-377."

Includes bibliographical references.

ISBN 0-8330-3833-8 (pbk. : alk. paper)

1. Airlift, Military—United States. 2. Deployment (Strategy) 3. Unified operations (Military science) 4. United States. Air Force—Transportation. I. Tripp, Robert S., 1944–

UC333.F73 2006
358.4'4'0973—dc22

2005024640

The RAND Corporation is a nonprofit research organization providing objective analysis and effective solutions that address the challenges facing the public and private sectors around the world. RAND's publications do not necessarily reflect the opinions of its research clients and sponsors.

Cover design by Stephen Bloodsworth

Published 2006 by the RAND Corporation
1776 Main Street, P.O. Box 2138, Santa Monica, CA 90407-2138
1200 South Hayes Street, Arlington, VA 22202-5050
4570 Fifth Avenue, Suite 600, Pittsburgh, PA 15213
RAND URL: http://www.rand.org/
To order RAND documents or to obtain additional information, contact
Distribution Services: Telephone: (310) 451-7002;
Fax: (310) 451-6915; Email: order@rand.org

Preface

This report examines options for improving the effectiveness and efficiency of intratheater airlift operations within the military joint end-to-end multimodal movement system. The intratheater system, which serves the needs of deploying, redeploying, and sustaining forces during contingency operations, is part of the airlift component of the joint movement system. This report discusses the application of an expanded strategies-to-tasks (STT) decision support framework to Central Command's (CENTCOM's) theater distribution planning and execution. We use the expanded STT framework to identify shortfalls and suggest, describe, and evaluate options for implementing improvements in current processes, organizations, doctrine, training, and systems. Specifically, we apply the framework to aid in improving planning and execution activities associated with developing airlift movement options in building and managing joint multimodal contingency movement networks. While the analysis centers on CENTCOM, the methodology and recommendations are relevant to other commands as well.

This work was conducted by the Resource Management Program of RAND Project AIR FORCE and was sponsored by the Commander of the U.S. Air Force, Central Command (CENTAF/CC). The research for this report was completed in October 2004.

This report should be of interest to combatant commanders and their staffs, mobility planners, logisticians, and planners throughout the Department of Defense (DoD), especially those in the Air Force and U.S. Transportation Command.

This report is one of a series of RAND reports that address agile combat support issues in implementing the Aerospace Expeditionary Force (AEF). Other publications issued as part of the larger project include:

- *Supporting Expeditionary Aerospace Forces: An Integrated Strategic Agile Combat Support Planning Framework*, Robert S. Tripp, Lionel A. Galway, Paul S. Killingsworth, Eric Peltz, Timothy L. Ramey, and John G. Drew (MR-1056-AF). This report describes an integrated combat support planning framework that may be used to evaluate support options on a continuing basis, particularly as technology, force structure, and threats change.
- *Supporting Expeditionary Aerospace Forces: New Agile Combat Support Postures*, Lionel A. Galway, Robert S. Tripp, Timothy L. Ramey, and John G. Drew (MR-1075-AF). This report describes how alternative resourcing of forward operating locations (FOLs) can support employment timelines for future AEF operations. It finds that rapid employment for combat requires some prepositioning of resources at FOLs.
- *Supporting Expeditionary Aerospace Forces: An Analysis of F-15 Avionics Options*, Eric Peltz, H. L. Shulman, Robert S. Tripp, Timothy L. Ramey, Randy King, and John G. Drew (MR-1174-AF). This report examines alternatives for meeting F-15 avionics maintenance requirements across a range of likely scenarios. The authors evaluate investments for new F-15 Avionics Intermediate Shop test equipment against several support options, including deploying maintenance capabilities with units, performing maintenance at forward support locations (FSLs), or performing all maintenance at the home station for deploying units.
- *Supporting Expeditionary Aerospace Forces: A Concept for Evolving to the Agile Combat Support/Mobility System of the Future*, Robert S. Tripp, Lionel A. Galway, Timothy L. Ramey, Mahyar A. Amouzegar, and Eric Peltz (MR-1179-AF). This report describes the vision for the agile combat support (ACS) system of the future based on individual commodity study results.

- *Supporting Expeditionary Aerospace Forces: Expanded Analysis of LANTIRN Options*, Amatzia Feinberg, H. L. Shulman, L. W. Miller, and Robert S. Tripp (MR-1225-AF). This report examines alternatives for meeting Low-Altitude Navigation and Targeting Infrared for Night (LANTIRN) support requirements for AEF operations. The authors evaluate investments for new LANTIRN test equipment against several support options, including deploying maintenance capabilities with units, performing maintenance at FSLs, or performing all maintenance at continental U.S. (CONUS) support hubs for deploying units.
- *Supporting Expeditionary Aerospace Forces: Lessons from the Air War over Serbia*, Amatzia Feinberg, Eric Peltz, James Leftwich, Robert S. Tripp, Mahyar A. Amouzegar, Russell Grunch, John Drew, Tom LaTourette, and Charles Robert Roll, Jr. (MR-1263-AF, not available to the general public). This report describes how the Air Force's ad hoc implementation of many elements of an expeditionary ACS structure to support the air war over Serbia offered opportunities to assess how well these elements actually supported combat operations and what the results imply for the configuration of the Air Force ACS structure. The findings support the efficacy of the emerging expeditionary ACS structural framework and the associated but still-evolving Air Force support strategies.
- *Supporting Expeditionary Aerospace Forces: Alternatives for Jet Engine Intermediate Maintenance*, Mahyar A. Amouzegar, Lionel A. Galway, and Amanda Geller (MR-1431-AF). This report evaluates the manner in which Jet Engine Intermediate Maintenance (JEIM) shops can best be configured to facilitate overseas deployments. The authors examine a number of JEIM supports options, which are distinguished primarily by the degree to which JEIM support is centralized or decentralized. See also *Engine Maintenance Systems Evaluation (En Masse): A User's Guide*, Mahyar A. Amouzegar and Lionel A. Galway (MR-1614-AF).
- *Supporting Expeditionary Aerospace Forces: Forward Support Location Options*, Tom LaTourrette, Donald Stevens, Amatzia Fein-

berg, John Gibson, and Robert S. Tripp (MR-1497-AF, not available to the general public).

- *A Combat Support Command and Control Architecture for Supporting the Expeditionary Aerospace Force*, James Leftwich, Robert S. Tripp, Amanda Geller, Patrick H. Mills, Tom LaTourrette, C. Robert Roll, Jr., Cauley Von Hoffman, and David Johansen (MR-1536-AF). This report outlines the framework for evaluating options for combat support execution planning and control. The analysis describes the combat support command and control operational architecture as it is now and as it should be in the future. It also describes the changes that must take place to achieve that future state.
- *Reconfiguring Footprint to Speed Expeditionary Aerospace Forces Deployment,* Lionel A. Galway, Mahyar A. Amouzegar, R. J. Hillestad, and Don Snyder (MR-1625-AF). This report develops an analysis framework—as a footprint configuration—to assist in devising and evaluating strategies for footprint reduction. The authors attempt to define footprint and to establish a way to monitor its reduction.
- *Analysis of Maintenance Forward Support Location Operations*, Amanda Geller, David George, Robert S. Tripp, Mahyar A. Amouzegar, C. Robert Roll, Jr. (MG-151-AF). This report discusses the conceptual development and recent implementation of maintenance forward support locations (also known as Centralized Intermediate Repair Facilities [CIRFs]) for the U.S. Air Force. The analysis focuses on the years leading up to and including the AF/IL CIRF test, which tested the operations of CIRFs in the European theater from September 2001 to February 2002.
- *Supporting Air and Space Expeditionary Forces: Lessons from Operation Enduring Freedom*, Robert S. Tripp, Kristin F. Lynch, John G. Drew, and Edward W. Chan (MR-1819-AF). This report describes the expeditionary ACS experiences during the war in Afghanistan and compares these experiences with those associated with Joint Task Force Nobel Anvil, the air war over Serbia. This report analyzes how ACS concepts were imple-

mented, compares current experiences to determine similarities and unique practices, and indicates how well the ACS framework performed during these contingency operations. From this analysis, the ACS framework may be updated to better support the AEF concept.

- *Supporting Air and Space Expeditionary Forces: A Methodology for Determining Air Force Deployment Requirements*, Don Snyder and Patrick Mills (MG-176-AF). This report outlines a methodology for determining manpower and equipment deployment requirements. It describes a prototype policy analysis support tool based on this methodology, the Strategic Tool for the Analysis of Required Transportation, that generates a list of capability units, called Unit Type Codes (UTCs), required to support a user-specified operation. The program also determines movement characteristics. A fully implemented tool based on this prototype should prove useful to the Air Force in both deliberate and crisis action planning.
- *Supporting Air and Space Expeditionary Forces: Lessons from Operation Iraqi Freedom*, Kristin F. Lynch, John G. Drew, Robert S. Tripp, and C. Robert Roll, Jr. (MG-193-AF). This report describes the expeditionary ACS experiences during the war in Iraq and compares these experiences with those associated with Joint Task Force Nobel Anvil, in Serbia, and Operation Enduring Freedom, in Afghanistan. This report analyzes how combat support performed, examines how ACS concepts were implemented in Iraq, and compares current experiences to determine similarities and unique practices. It also indicates how well the ACS framework performed during these contingency operations.
- *Supporting Air and Space Expeditionary Forces: Analysis of Combat Support Basing Options*, Mahyar A. Amouzegar, Robert S. Tripp, Ronald G. McGarvey, Edward W. Chan, and C. Robert Roll, Jr. (MG-261-AF). This report evaluates a set of global FSL basing and transportation options for storing war reserve materiel. The authors present an analytical framework that can be used to evaluate alternative FSL options. A central

component of the authors' framework is an optimization model that allows a user to select the best mix of land-based and sea-based FSLs for a given set of operational scenarios, thereby reducing costs while supporting a range of contingency operations.

RAND Project AIR FORCE

RAND Project AIR FORCE (PAF), a division of the RAND Corporation, is the U.S. Air Force's federally funded research and development center for studies and analyses. PAF provides the Air Force with independent analyses of policy alternatives affecting the development, employment, combat readiness, and support of current and future aerospace forces. Research is conducted in four programs: Aerospace Force Development; Manpower, Personnel and Training; Resource Management; and Strategy and Doctrine.

Additional information about PAF is available on our Web site at http://www.rand.org/paf.

Contents

Figures

Tables

Summary

In summer 2003, after the major combat phase of Operation Iraqi Freedom, the commander of Air Force forces (COMAFFOR) for U.S. Central Command (CENTCOM), Lieutenant General Walter Buchanan, recognized the need to undertake a fundamental reexamination of the Theater Distribution System (TDS). The COMAFFOR and his Director of Mobility Forces (DIRMOBFOR) manage the Air Mobility Division (AMD)—part of the Coalition Air and Space Operations Center (CAOC)—and are responsible for planning and executing the airlift component of the TDS. The AMD also provides common user airlift services, in addition to other responsibilities, to U.S. and coalition forces within the area of responsibility (AOR). The COMAFFOR noticed several problems associated with the planning and execution of airlift including:

- A backlog of cargo at aerial ports of debarkation/embarkation (APODs/APOEs)
- Incomplete visibility of cargo within the TDS
- Information system connectivity issues with air terminal operations centers operated by the components
- Apparent inefficient use of airlift resources
- A lack of discipline in requesting airlift support
- Perception of inadequate support for intratheater airlift resources.

During the same time period, the Secretary of Defense assigned deployment process ownership to Joint Forces Command and distribution process ownership to U.S. Transportation Command (USTRANSCOM) in an attempt to improve those processes and address problems that arose in Operation Allied Force, Operation Enduring Freedom, and Operation Iraqi Freedom. As part of executing its new responsibilities, USTRANSCOM, with the consent of the Commander of CENTCOM, created a CENTCOM Deployed Distribution Operations Center (C-DDOC) in the AOR. The C-DDOC works for the CENTCOM J4[1] and was created to improve the joint multimodal TDS and better integrate it with the joint multimodal intertheater movement system. The C-DDOC absorbed the personnel and duties of the Joint Movement Center (JMC) when it deployed to the AOR.

Analytic Approach

In light of today's changing security environment, the objective of our analysis was to evaluate options for improving CENTCOM's theater airlift planning and execution to support the joint expeditionary contingency operations. This research should provide a solid foundation for the Defense Department (DoD) to use in addressing issues in the theater airlift planning and execution system observed in recent contingencies. We use an expanded strategies-to-tasks (STT) framework as a "lens" for evaluating intra- and intertheater movement planning and execution processes. This expanded framework incorporates resource allocation processes and constraints in movement planning and execution activities. It also describes how movement resources and processes can be related to operational effects. Finally, this analytic framework recognizes that no optimal solution

[1] As defined in Air Force Doctrine Document–2, *Organization and Employment for Air and Space Operations,* the Air Force terminology used by the authors identifies organizations/responsibilities. The A/J3 is the Operations Directorate; the A/J4 is the Logistics Directorate; and A/J5 is the Plans Directorate (with A standing for "Air Force" and J standing for "Joint").

exists for configuring contingency movement networks. Rather, the network is derived from a set of choices on how limited movement resources can be used (see pp. 6–9).

The first step was to apply the STT framework to theater airlift planning and execution to derive its operational objectives. We step from the National Security Strategy down through National Military Strategy, National Military Objectives, and relevant campaign objectives to reach these. We incorporate extensive input from subject-matter experts at site visits to CENTCOM, CENTAF, ARCENT, USTRANSCOM, AMC, Expeditionary Mobility Task Forces (EMTFs), and the Air Staff.[2] We then expand the STT framework by applying a useful resource management framework from prior RAND research and a more generic closed-loop planning construct to complete our theater airlift planning and execution framework.

Using this expanded STT framework, we identify *supply-side* processes associated with planning, replanning, and executing common user contingency airlift operations within the combatant commander's (COCOM's) AOR and coordinating these activities within the end-to-end joint movement systems. We identify *demand-side* processes associated with common user contingency airlift operations. Finally, we identify *integrator* processes associated with allocating scarce movement resources to those needs with the highest COCOM priorities.

We use this expanded framework to examine the AS-IS[3] set of processes, organizations, doctrine, training, and systems. We identify disconnects and missing processes against those identified as being necessary in the expanded STT framework. We then identify TO-BE options that can be developed to address disconnects and missing processes. Finally, we evaluate the effectiveness and efficiency of the TO-BE options.

---

[2] For a complete list of organizations involved in this research, see Appendix A.

[3] When this study began, CENTCOM had theater airlift planning and execution processes in place as outlined in doctrine. As such, the term *AS-IS* refers to both the CENTCOM theater airlift planning and execution processes and the processes outlined in doctrine.

AS-IS Theater Airlift Planning and Execution Shortfalls

Using the expanded STT framework with the closed-loop planning construct and the analysis of the theater airlift planning and execution system, we documented the AS-IS theater airlift planning and execution system and compared it with the attributes derived from applying the expanded STT framework to identify existing shortfalls in process, organization, doctrine, training, and communications and systems (see pp. 27–58). The shortfalls are shown in Figure S.1.

Theater Airlift Planning and Execution TO-BE Improvement Options

Our analyses suggest improvements in process, organization, doctrine, training, and information systems (see pp. 59–81).

Figure S.1
AS-IS Theater Airlift Planning and Execution Shortfalls

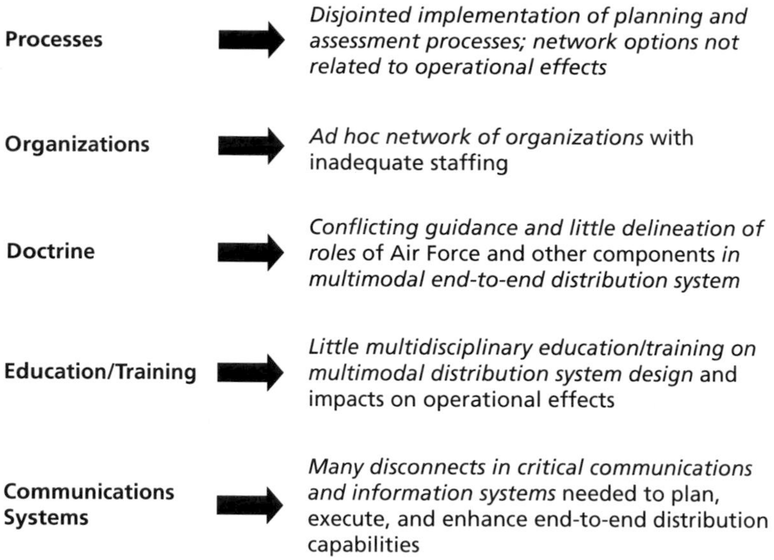
Processes
Disjointed implementation of planning and assessment processes; network options not related to operational effects
Organizations
Ad hoc network of organizations with inadequate staffing
Doctrine
Conflicting guidance and little delineation of roles of Air Force and other components in multimodal end-to-end distribution system
Education/Training
Little multidisciplinary education/training on multimodal distribution system design and impacts on operational effects
Communications Systems
Many disconnects in critical communications and information systems needed to plan, execute, and enhance end-to-end distribution capabilities

RAND *MG377-S.1*

Process and Organization Improvements

Using the analytic approach described above, we generated two organizational options: modifying existing process assignments to the J3/5 and J4 using the expanded STT framework and creating a new line organization responsible for all end-to-end movement (see pp. 59–75).

We first evaluated realigning processes, doctrine, organizations, training, and communications and information systems among existing organizations responsible for planning and executing the joint multimodal end-to-end movement system. This option uses the expanded STT framework and assigns responsibilities for movement planning and execution to joint and component organizations consistent with the demand, supply, and integrator roles.

Building on modified assignment of responsibilities, we then evaluated creating a Deployment and Distribution Movement organization responsible for planning and executing the TDS in conjunction with the intertheater movements system. The director could be dual-hatted with USTRANSCOM to better integrate intertheater movement planning and execution requirements.

We then evaluated using personnel in CONUS, through reachback, to provide some of the realigned products and services. Using reachback for assistance in tactical-level planning shows promise of better effectiveness and efficiency with a reduced footprint. Reachback support could enhance routing and scheduling of airlift within the multimodal movement system.

Finally, we looked at creating a Joint Theater Logistics Commander (JTLC) organization that would be responsible for planning and executing the TDS with the intertheater movements system. USTRANSCOM would retain intertheater movement planning and execution processes in this case.

Our analysis suggests that the separation of demand, supply, and integrator responsibilities can strengthen integrated movement planning. Also, the adoption of a closed-loop planning and execution process that focuses on the trade-offs in effectiveness and efficiency of alternative network options has promise for improving decisions on network design. Metrics that show demand side and supply side trade-offs should be used routinely to reinforce the notion that there is no one right answer, but rather a set of options with greater or lesser effectiveness and costs.

The placement of all joint movement forecasting and prioritization in a demand-side organization focuses attention on the entire range of movement requirements. This single focal point for movement requirements and priorities also fixes one of the main problems in the AS-IS process—that of conflicting movement guidance between sustainment and deployment/redeployment requirements.

The placement of strategic-level (for example, network design and determining key nodes and transshipment points) and operational-level planning functions (for example, changing routes within a specified network or adding capacity) for intratheater distribution within a single joint organization responsible for developing and assessing the effectiveness and efficiency of network options also clarifies responsibilities. While tactical level planning—for example, determining specific airlift routes and schedules—can be enhanced, our analysis indicates that greater payoffs lie in improving the strategic- and operational-level planning processes.

Enhancing existing processes by modifying assignment of responsibility using an expanded STT framework appears to be the least intrusive option. To gain benefit from changes in assignment of responsibilities among existing organizations, this option should be accompanied by improvements in existing processes, changes in doctrine, and training enhancements. Investment in communications and information systems also could be helpful.

Based on our analysis, using an expanded resource allocation STT framework to separate supply, demand, and integrator processes and assigning them to existing organization to improve effectiveness and efficiency should be implemented. Applying the expanded STT framework should help the J4 do his or her job. Either organizational option will work. Then, a thorough review of possible reachback options should be completed and the JTLC option can be explored in depth.

Modifying process assignments using an expanded STT framework is appropriate for several reasons. It is consistent with time-tested doctrine that has guided contingency operations for many decades. This basic doctrine calls for COCOMs to develop and execute contingency plans subject to oversight by the Secretary of Defense. Using this doctrine, COCOMs are responsible for employing forces. The services are responsible for providing forces. Using our expanded framework, the COCOM is a demand organization. The components and the specified joint commands, such as USTRANSCOM, are supply-side organizations. Thus, at the highest level, the Secretary of Defense is the integrator among COCOM demands and component and specified joint command suppliers.

Modifying existing process assignments also ensures that COCOM priorities are met by assigning operational control of intratheater resources to the COCOM and having access to agreed-on or arbitrated allocations of intertheater movement resources. Given adequate planning and guidance, this option supports agility in meeting dynamically changing battlefield conditions by having in-theater movement resources under the control of the COCOM.

This option also strengthens joint strategic- and operational-level planning and assessment while leaving tactical planning and execution responsibilities in the hands of the components, preserving unity of command.

Using the expanded STT framework to realign processes should be relatively easy because it deals with changing processes and clear assignment of responsibilities to existing organizations. This option does not create a new hybrid organization that the Joint Theater Logistics Command would create. It also does not extend centralized execution. This option would assign intra- and intertheater movement planning and execution responsibilities to standing organizations in each COCOM, at USTRANSCOM, and within each component.

To improve airlift planning and execution within the joint multimodal end-to-end movement system, the following actions are needed to modify and enhance process assignments within existing organizations.

- Airlift planning expertise within the COMAFFOR A3/5 needs enhancing. We estimate that two additional airlift planners are needed for this purpose for each COMAFFOR.
- Assessment capabilities in the AMD should be created. An Assessment Cell should be created and staffed with a small analysis team, potentially through reachback. We estimate that two airlift planners are needed for this process in each AMD/AMOCC.
- Supply-side network planning responsibilities should be separated from demand-side planning responsibilities. The J4 should be established as the integrated COCOM movements planning and execution supply-side focal point. J4 Movement System Planning (currently in X-DDOCs) should be separated from Assessment and Allocation responsibilities (JMC responsibility as outlined in Joint Publications). This move does not affect staffing requirements on its face, but the J4 Movement System Planning Organization would require some of the best-educated and trained airlift planners. Some of these planning functions could be supported via reachback to the COMAFFOR A3/5 enhanced staff and to the TACC and USTRANSCOM DDOC.
- A J3 organization needs to be established and staffed to perform integrated requirements forecasts and guidance (demand-side). We estimate that six total slots would be needed to support this process (three of which would be Air Force slots). Embedding a group of J4 planners within the J3 organization could allow J3 planners to focus on operations while the embedded J4 planners focus on prioritizing movement requirements.
- Reachback support to the TACC for tactical planning can result in lower personnel requirements and reduced footprint. Communications appear adequate to support these processes. Specific reachback responsibilities and organizations need to be defined.

Doctrine, Training, Communications, and Information System Changes

To effectively implement either the improved interfaces option or the creation of the Deployment and Distribution Movement Organization, doctrine must be revised and enforced. For example, Joint Publications 4-0, 4-01.3, and 5 will have to be revised just to name a few (see pp. 75–80). Any doctrine that outlines responsibilities for the A3/5, AMD, TACC, X-DDOCs, J3, or J4 will have to be revised. These documents will need to be changed to reflect process and organizational discussions as outlined above.

The improved interfaces and the Deployment and Distribution Movement Organization options also have significant training implications for each of the components, and the Air Force in particular. Each COMAFFOR, as well as COCOMs, USTRANSCOM, and the 18th Air Force, should be provided with trained personnel who are educated and experienced in multimodal movement planning and execution and STT methods and tools. The Air Force may need to invest in multimodal training and establish educational identifiers to track training. STT education could be provided through Air Force continuing education in such venues as the Air Mobility Warfare Center (AMWC) and other Air Force schools. For example, the Contingency War Planners Course and the Joint Air Operations Planning Course could be used to increase awareness. Log 399 could provide immersion for anyone involved in J3 demand generation. The Army's Transportation School could be used for a multimodal planning and execution course aimed at the joint end-to-end movement system development in contingency environments. In addition, graduate courses could be developed at the Air Force Institute of Technology (AFIT).

Communications and information system connectivity will also need to be enhanced. Currently, communications and information system disconnects exist between the AMD and the component operational units and air terminal operations centers (ATOCs) operated by the different components. Different systems and communications architectures carry information on airlift cargo, requirements, eligibility, and status. Some systems use the NIPRNET, some the SIPRNET. These disconnects make it difficult to determine requirements for airlift and effectively schedule it to meet the needs of component operational and support units. A common systems architecture is needed. Perhaps the Global Air Transportation Execution

System (GATES) could be the common system used for all movement requirements. Having Radio Frequency Identification and Detection (RFID) tags read directly into GATES could solve some of the asset visibility issues.

Acknowledgments

This work would not have been possible without the support of many individuals. First, we thank our sponsor, Lt Gen Walter Buchanan, Commander, U.S. Air Force Central Command and Combined Forces Air Component Commander (CFACC). We also thank Maj Gen Robert Elder, Deputy CFACC, for his help in shaping this report, providing access to his staff, and facilitating our travel within the Central Command area of responsibility. Brigadier General David Stringer, former Deputy J4, European Command, and currently commander of the Arnold Engineering Development Center, provided a thorough review of this report and was instrumental in improving it.

Briefings of this work were given to a number of people. Their comments and insights have helped shape the analysis. We thank Lt Gen William Welser, Commander, 18th Air Force; Lt Gen Donald Wetekam, Deputy Chief of Staff, Installations and Logistics; Lt Gen Stephen Croker, U.S. Air Force (Ret.), Air Force Senior Mentor; Lt Gen Charles Heflebower, U.S. Air Force (Ret.), Air Force Senior Mentor; Admiral Christopher Ames, U.S. Transportation Command (USTRANSCOM)/J5; Maj Gen Robert Dail, USA, USTRANS-COM/J3; Maj Gen James Hawkins, C-DDOC, 18AF/CV; Maj Gen Craig Rasmussen, AF/ILG; Maj Gen Norman Seip, CENTAF CAOC Deputy CFACC; Brig Gen Rick Ash, DIRMOBFOR; Brig Gen Allen Peck, CENTAF/CV; Brig Gen Rick Perraut, 15 EMTF/CC, DIRMOBFOR; Brig Gen Paul Selva, TACC/CC; Brig Gen Bobby Wilkes, 21 EMTF/CC; Brig Gen Mark Zamzow, CENTAF-

AUAB CAOC DIRMOBFOR; and Col Mike Morabito, Commander, Air Force Logistics Management Agency, and his AFLMA staff.

In particular, we would like to thank Col Bruce Busler, Deputy DIRMOBFOR and AMD Director, 15 EMTF/CV, for his help with theater visits, access, and his many comments and feedback.

Many people contributed to our visits to the CENTCOM AOR. For access to C-DDOC and briefing us on its mission, we thank Brig Gen Levasseur, Commander, C-DDOC, and his staff. We thank Col Cathy Robertello, JMC Director, for access to the JMC and her thoughtful comments and feedback. At the AMD, we thank Col Brooks Bash, Deputy DIRMOBFOR, and Director of AMD. We also thank Lt Col Hans Petry and Maj Matt Lacy for guiding us through the AMD. In the AFFOR/A4, we thank Col Michael Butler, A4, and Maj David Paynter for helping coordinate our visit; Maj Kirk Patterson and SMSgt Frederic Hale for informative and illuminating conversations on communications and information systems. From DynCorp, we thank Mr. Al Pianalto. We also thank Lt Col David Frazee, Maj Azad Keval, SMSgt Gary Shirley, Mr. Bill Lemay, and Mr. Rick Irby.

A number of individuals at USTRANSCOM and Air Mobility Command (AMC) helped us with visits and interviews. At USTRANSCOM, we thank Mr. George Raney, and his staff, including Maj Bobbie Leyes for helping to arrange and facilitate our visit, and Mr. Stu Draper of MITRE. At AMC headquarters, we thank Col Henry Haisch, TACC/XON, for insight and feedback, as well as his staff; Col Paul Curtis, AMC/A43, and his staff, particularly Mr. Don Siegel and Maj Dan Bradley; Lt Col Jane Clarke, AMC/A5, and her staff, particularly Mr. Ed Acosta, AMC/A5, and Mr. Dave Merrill, AMC/A59. We also thank Lt Col Bob Eubanks, AMC/A31, Mr. Mo Verling, AMC/A38IP, Mr. Gary Little, AMC/A38B, and Mr. Roger Beumann, AMC/A66C.

On the Air Staff, we thank Col Chris O'Hara, AF/ILGS, for comment, review, and support in gathering data.

From AFIT, we thank Maj Victor Wiley, Maj John Bell, and Maj Jeff Brown.

At RAND, we benefited from discussions and comments form a number of individuals. Among those, we thank (in alphabetical order): Mahyar Amouzegar, Ed Chan, Lionel Galway, Jim Masters, Ray Pyles, and Tim Ramey. We especially thank David Thaler and Carl Rhodes for their thorough review of this report. Their reviews helped shape this report into its final, improved form.

All these have contributed to this research; we assume responsibility for any errors.

Abbreviations

ACS	Agile Combat Support
AE	Aeromedical Evacuation
AECT	Aeromedical Evacuation Control Team
ALCT	Airlift Control Team
AMC	Air Mobility Command
AMCT	Air Mobility Control Team
AMD	Air Mobility Division
AME	Air Mobility Element
ANG	Air National Guard
AOC	Air and Space Operations Center
AOR	Area of responsibility
APOD	Aerial port of debarkation
APOE	Aerial port of embarkation
ARCENT	U.S. Army, Central Command
ATO	Air tasking order
ATOC	Air terminal operations center
CAOC	Combined Air and Space Operations Center
CC	Commander
CDC	Corps Distribution Center

C-DDOC	CENTCOM Deployment and Distribution Operations Center
CENTAF	U.S. Air Force, Central Command
CENTCOM	U.S. Central Command
CFACC	Combined Forces Air Component Commander
CFLCC	Combined Forces Land Component Commander
CJTF	Combined Joint Task Force
CIRF	Centralized Intermediate Repair Facility
CMOS	Cargo Movement Operations System
COCOM	Combatant commander
COMAFFOR	Commander of Air Force Forces
COMALF	Commander of Airlift Forces
CONOP	Concept of operation
CV	Vice commander
DDOC	Deployment and Distribution Operations Center
DIRMOBFOR	Director of Mobility Forces
DoD	Department of Defense
Dyna-METRIC	Dynamic Multiechelon Technique for Recoverable Item Control
EMTF	Expeditionary Mobility Task Force
EUCOM	U.S. European Command
FOL	Forward operating location
FRAG	Fragmentation order
GATES	Global Air Transportation Execution System
ISR	Intelligence, Surveillance, and Reconnaissance

JFACC	Joint Forces Air Component Commander
JMC	Joint Movement Center
JMMS	Joint Multimodal Movement System
JOPES	Joint Operations Planning and Execution System
JTF	Joint task force
JTLC	Joint Theater Logistics Commander
LOI	Letter of Instruction
METRIC	Multiechelon Technique for Recoverable Item Control
MOE	Measure of effectiveness
MTW	Major theater war
NIPRNET	Nonsecure Internet Protocol Router Network
PACOM	U.S. Pacific Command
PID	Plan identification designator
RDD	Required delivery date
RFID	Radio Frequency Identification and Detection
RSP	Readiness Spares Package
SAS	Senior Air Service
SIPRNET	Secret Internet Protocol Router Network
SOF	Special operations forces
STT	Strategies-to-tasks
SWA	Southwest Asia
TACC	Tanker Airlift Control Center
TCACIS	Transportation Coordinator's Automated Cargo Information System

TCAIMS	Transportation Coordinator's Automated Information for Movements System
TDS	Theater Distribution System
TOA	Table of Allowance
TPFDD	Time-Phased Force Deployment Data
TWCF	Transportation Working Capital Fund
USTRANSCOM	U.S. Transportation Command
ULN	Unit line number
UTC	Unit Type Code
WRM	War reserve materiel

CHAPTER ONE

Introduction

Airlift planning and execution, part of the Theater Distribution System (TDS), are vital parts of combat support execution planning and control. In today's security environment, combat forces are expected to react quickly to any national security issue with a tailored, sustainable force. An operation's success relies on the movement of personnel and equipment. Without a reliable movement system, deployment can be delayed and sustainment can be hindered. This report examines options for improving the effectiveness and efficiency of intratheater airlift operations within the military joint end-to-end multimodal movement system that serves the needs of deploying, redeploying, and sustaining forces during contingency operations.

Motivation for the Analysis

The United States has had military presence in the Central Command (CENTCOM) area of responsibility (AOR) almost continuously since the demise of the Soviet Union in 1990. During this time, U.S. military forces undertook four major operations. In 1991, Operation Desert Storm brought more than 500,000 U.S. military personnel to CENTCOM (USAF, 1993, Vol. V, Part I, Table 19, p. 61). As many as 150 C-130s were in theater, and they flew more than 1,200 tactical airlift missions (USAF, 1993, Vol. V, Part I, Table 21, pp. 65 and 250). An Air Force Brigadier General was designated the CENTCOM Commander of Airlift Forces (COMALF) (USAF,

1993, Vol. III, Part I, p. 147).[1] The COMALF provided command and control of theater airlift forces through the Airlift Control Center (similar to today's Air Mobility Division [AMD] in the Air and Space Operations Center [AOC]). During Desert Storm, theater distribution problems arose, including the arrival of combat forces before adequate combat support and intratheater movement capabilities were established and poor in-transit visibility.[2]

After Operation Desert Storm, operations and logistical requirements were relatively steady for the decade prior to Operation Enduring Freedom. In 2001, just prior to Operation Enduring Freedom, TDS in the CENTCOM AOR consisted of command and control of four C-130s in support of Operation Southern Watch.[3] The Air Force maintained responsibility over the CENTCOM TDS. During this time, operations did not require a mature command structure for airlift or a mature communications infrastructure.

Initiated in October 2001, Operation Enduring Freedom began the largest U.S. military mobility operation since Operation Desert Storm. This approximately tripled the Southern Watch presence already in the AOR (Lynch et al., 2005). CENTCOM delegated responsibility for the TDS—the planning and execution of all movements of materiel and personnel within the AOR by land (trucks and rail), sea (ships and barges), and air—and for the Joint Movement Center (JMC) to the Air Force. Although typically an Army responsibility, TDS responsibility can, according to joint doctrine, be appointed to any service based on "either the dominant-user or the most-capable-service concept" (Joint Chiefs of Staff, 1996, p. v). During Operation Enduring Freedom, initial responsibility was given to the Air Force with the understanding that the Army would

---

[1] Under U.S. Air Force Lieutenant General Charles Horner, who was dual-hatted as the CENTAF/CC and CENTCOM/JFACC.

[2] See USAF (1993, Vol. III, Part I, p. 144) for a summary.

[3] Operation Southern Watch enforced the no-fly zone in southern Iraq.

assume responsibility once ground forces were engaged.[4] Twenty-four tactical aircraft flew 2,700 tactical airlift sorties.[5] As in Operation Desert Storm, TDS problems emerged. Large backlogs of cargo developed at transshipment points in the AOR during Operation Enduring Freedom and standard air routes (STARs) were not established early enough to meet TDS needs (Tripp et al., 2004).

Operation Iraqi Freedom, which started in March 2003, saw the deployment of approximately 200,000 U.S. servicemembers to the CENTCOM AOR. Demand for supplies increased more than 300 times over a period of just a few months.[6] More than 120 C-130s began operating theater missions.[7]

After the conclusion of major combat operations in Iraq, TDS problems continued. The commander of Air Force forces (COMAFFOR) for CENTCOM observed several symptoms of problems associated with TDS, including:

- Difficulty in predicting cargo requirements
- Difficulty in configuring, reconfiguring, basing, and sizing TDS airlift
- Confusion on appropriate metrics to judge airlift effectiveness
- Appearance of incomplete coordination of movement modes in meeting TDS needs
- Incomplete visibility of cargo within the TDS
- Artificial separation of strategic movements system from TDS
- Restriction of strategic airlift assets for intratheater use in early phases of conflict
- Inefficient use of intratheater airlift assets.[8]

---

[4] The combatant commander stipulated that TDS responsibility would transfer to the Combined Forces Land Component Commander (CFLCC) once ground forces were engaged, but this had not occurred more than 18 months after Operation Enduring Freedom began.

[5] According to the Air Force news Web site and CNN Web site.

[6] Analysis of GATES data, January to October 2003.

[7] Interview with CENTCOM Air Mobility Division, April 2004.

[8] Conversations with DCFACC for CENTCOM and Deputy DIRMOBFOR and Director of the AMD.

Many of these symptoms were recognized during Operations Desert Storm, Enduring Freedom, and again during Iraqi Freedom. As a result of these continuing issues, in August 2003, the COMAFFOR asked RAND Project AIR FORCE to analyze options for improving the effectiveness and efficiency of intratheater airlift operations in southwest Asia.

Focus and Scope of the Analysis

Requirements for the airlift portion of the joint movement system are met through intertheater and intratheater resources. These airlift needs can be met by military or commercial capabilities depending on threat conditions, cargo characteristics, and other factors, as deemed appropriate by military planners. Experience after Operation Iraqi Freedom indicates that systemic problems exist in TDS planning and execution and that a thorough examination of existing processes, doctrine, organization, training, and systems is needed. In evaluating TDS, options considered must recognize that airlift operations are part of an integrated end-to-end multimodal distribution system. In addition, airlift operations in contingency operations must be flexible and responsive to rapidly changing needs on the battlefield.

Table 1.1 illustrates the relationships among contingency movement planning and execution capabilities (on the left-hand side of the table) and joint expeditionary combat support effects (on the right).[9] For example, to tailor force and combat support packages needed to achieve desired operational effects, a capability is needed to estimate needed movement requirements to meet the specific beddown and mission requirements, such as tailored Unit Type Codes (UTCs). To employ forces quickly, a capability is needed to configure the movement network quickly to deliver combat and support

[9] Joint expeditionary combat support effects are from Tripp et al. (2000).

Table 1.1
Integrated Movement Planning and Execution Capabilities Create Joint Combat Support Effects

Joint Expeditionary Combat Support Effects	Contingency Movement Planning/ Execution Capabilities
Rapidly tailor force and support packages to achieve desired operational effects	Estimate inter- and intratheater movement requirements for selected force and support package options
Deploy rapidly	Facilitate rapid TPFDD development Assess feasibility, cost, and time of deliveries Configure inter- and intratheater movement system Determine FOL beddown capabilities for force packages
Employ quickly	Configure movement network rapidly to meet employment timelines and sutainment needs
Balance deployment and sustainment resource allocations	Apply resources to meet deployment and sustainment movement needs Assess network performance and reconfigure as needed
Allocate scarce resources to where they are needed most	Determine impacts of allocating scarce resources to various JTFs and prioritize allocations to users
Adapt to changes quickly	Indicate when movement performance deviates from desired state and implement get-well plans

NOTE: TPFDD = Time-Phased Force Deployment Data; FOL = forward operating location; JTF = joint task force.

resources needed to conduct initial and follow-on employment operations.

The relationships between these movement planning and execution capabilities and joint combat support effects are receiving much attention and are beginning to be understood by both the operations and combat support communities. Over time, we need to extend the thinking shown in Table 1.1 to relate movement planning and execution processes to achieving joint operational effects by supporting the combatant commander's (COCOM's) campaign plan.

For the purpose of this report, strategic and theater airlift planning and execution activities, such as the repetitive planning, executing, and replanning of airlift resources to meet COCOM needs in his or her area of responsibility (AOR), are considered. These activities include:

- Developing an airlift network, including identification of nodes and different types of routes—for example, demand-based and frequency-based routes.
- Estimating the aircraft and crews needed as functions of basing options available.
- Deployment of communications and information systems needed to manage and control airlift operations.
- Deployment and sustainment of resources needed to run air terminal operations.
- Combat support resources needed to house deployed airlift operations at forward and main operating bases.

Intratheater airlift operations include:

- Onward movement of deploying forces from APODs within the AOR to airfields at or near their initial deployment sites.
- Redeployment of units from field locations to AOR APOEs for return to CONUS or other home stations.
- Movement of forces within the AOR from one area to another as dictated by battlefield necessities.
- Movement of sustainment cargo and replacement personnel.
- Movement of soldiers to APOEs for authorized leave within their tours of duty within the AOR.
- Movement of war reserve materiel (WRM) within the AOR as necessary to establish forward operating bases and to redeploy those assets to WRM sites within the AOR for reconstitution.

Analytic Approach

An effective and efficient TDS is necessary to support a military force able to react quickly to any national security issue. The purpose of this report is to present a framework for approaching theater airlift planning and execution in the context of the global mobility system and recommend policy options to improve its performance. By understanding movement planning and execution processes, we are able to suggest improvements in assignment of responsibilities, training and education, and systems and tools. To this end, we use an expanded strategies-to-tasks (STT) framework (see Appendix B) as a "lens" for evaluating intra- and intertheater movement planning and execution processes. These processes include assessing demand requirements, establishing beddown sites for airlifters, establishing transshipment points, determining fleet sizes and types of aircraft to meet demands, establishing routes and schedules, deploying communications and information systems, terminating and redeploying resources when contingency operations end, and integrating the intra- and intertheater airlift system into the end-to-end joint multimodal movement system. Expanding the basic STT framework, we incorporate resource allocation processes and constraints into movement planning and execution activities. We also describe how movement resources and processes can be related to operational effects. Finally, using this framework for analysis, we recognize that no optimal solution exists for configuring contingency movement networks. Rather, the network is derived from a set of choices on how limited movement resources can be used.

Using this expanded STT framework, we identify *supply-side* processes associated with planning, replanning, and executing common user contingency airlift operations within the COCOM AOR and coordinating these activities within the end-to-end joint movement systems. We identify *demand-side* processes associated with common user contingency airlift operations, and we identify *integrator* processes associated with allocating scarce movement resources to those needs with the highest COCOM priorities.

We use this expanded STT framework to examine the AS-IS[10] set of processes, organizations, doctrine, training, and systems. We

---

[10] When this study began, CENTCOM had theater airlift planning and execution processes in place as outlined in doctrine. As such, the term *AS-IS* refers to both the CENTCOM theater airlift planning and execution processes and the processes outlined in doctrine.

identify disconnects and missing processes by comparing the AS-IS against those processes that are identified as being necessary theater airlift planning and execution processes in the expanded STT framework. We then identify TO-BE options that can be developed to address disconnects and missing processes.

Finally, we evaluate the effectiveness and efficiency of the TO-BE options. See Figure 1.1 for a diagram of our analytic approach.

In addition to applying the expanded STT framework to theater airlift planning and execution, we document the AS-IS theater airlift planning and execution system. The research team began by evaluating how CENTCOM currently plans and executes theater airlift operations. During site visits, we interviewed individuals involved in

Figure 1.1
Analytic Approach

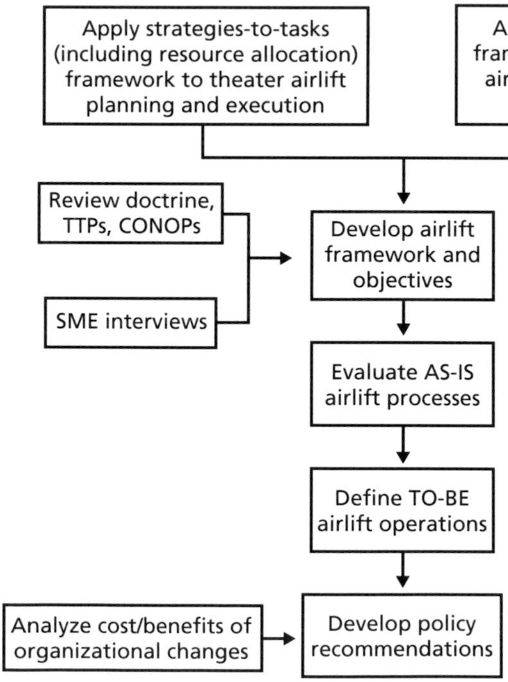

RAND *MG377-1.1*

airlift planning and execution.[11] The team reviewed internal memorandums and Air Force and joint doctrine, manuals, instructions, and concepts of operations (CONOPs) and described the processes and organizational responsibilities derived from the documents, interviews, and analyses of recent contingencies.

We have worked with Air Force, other Service, and joint COCOM stakeholders in conducting this research. Each organization openly and candidly discussed issues associated with TDS planning and execution from their vantage points. Each was interested in helping us address options for improving TDS options and ensuring that our results could be implemented. Our aim is to improve theater airlift planning and execution, but it is so enmeshed with the Joint Multimodal Movement System (JMMS) that our framework and some policy recommendations reach beyond theater airlift.

Related Activities

Any analysis should be considered in context with other ongoing initiatives that may have an impact on outcomes and potential implementation actions associated with the analysis. These ongoing initiatives include actions by the Secretary of Defense, in September 2003, to assign ownership of the military distribution process to U.S. Transportation Command (USTRANSCOM) and ownership of the deployment process to Joint Forces Command (denoted as the Distribution Process Owner and Deployment Process Owner, respectively). The intent of these assignments is to give responsibility and authority to *one* agency for developing and improving processes that rely on *many* organizations to execute the process. For example in distribution activities, cargo preparation, movement, and receipts are generally done by different organizations, at different echelons, in different services.

---

[11] See Appendix A for a complete list of organizations that participated in this research.

The analysis needs to be consistent with joint vision and doctrine as captured in the Air Force Agile Combat Support, Global Mobility, Command, Control, Communications, Computers, Intelligence, Surveillance, and Reconnaissance (C4ISR) CONOPs that will guide Air Force operations in the future. In addition, policymakers need to understand the impacts for TDS resulting from the Army's initiatives to create smaller, more mobile units, including the unit of action. The Army has also sponsored an analysis of the intratheater distribution system as a result of problems that they experienced with supply movements keeping pace with the rapid movement of combat units in Operation Iraqi Freedom. This analysis is being undertaken by the RAND Arroyo Center at the request of the Army Deputy Chief of Staff for Logistics. All improvement options need to be considered within the context of these and other ongoing initiatives.

Organization of This Report

In Chapter Two, we apply the expanded STT and closed-loop frameworks to theater airlift planning and execution from which we derive theater airlift planning and execution requirements, including discussions on resource management and planning processes for theater airlift planning and execution. Chapter Three discusses problems with the AS-IS CENTCOM TDS. Chapter Four suggests revisions to CENTCOM's theater airlift planning and execution process, with organizational, doctrinal, and training changes to support the revisions. Chapter Five contains our conclusions and recommendations. Appendix A lists the organizations contributing to this analysis. Appendix B presents the basic STT and closed-loop frameworks. Appendix C is an illustration of the closed-loop planning and execution process. Appendix D is an illustrative example of how reachback can be used in the AMD, and Appendix E is the Reachback Support Decision Tree used in reachback decisionmaking. Appendix F outlines CENTCOM's evolved intratheater airlift planning process.

CHAPTER TWO

Strategies-to-Tasks and Closed-Loop Planning Applied to Theater Airlift

The STT framework was developed at RAND during the late 1980s[1] and has been widely applied in the Department of Defense (DoD) to aid in strategy development, campaign analysis, and modernization planning.[2] The framework has proven to be a useful approach for providing intellectual structure to ill-defined or complex problems. Working through the STT hierarchy can help identify areas where new capabilities are needed, clarify responsibilities among actors contributing to accomplishing a task or an objective, and place into a common framework the contributions of multiple entities and organizations working to achieve some common objective. In this analysis, we use an expanded STT framework to show how combat support elements, or more specifically movement capabilities, can be related to task-organized operational elements used to create desired joint operational effects by supporting the COCOM's campaign plan.

A closed-loop assessment and feedback process[3] is a concept that has been well understood in operational planning and has been the topic of operational planning doctrine for many years (Boyd, 1987). This process can inform operational planners of how the performance

[1] See Kent (1989) and Thaler (1993).

[2] Internal examples are Lewis et al. (1999) and Niblack, Szayna, and Bordeaux (1996). Outside of RAND, the framework is in use by the Air Force, the Army, and elements of the Joint Staff.

[3] A *closed-loop* process takes the output and uses it as an input for the next iteration of the process.

of a particular combat support process affects operational capability. The closed-loop process can be applied to such critical tasks as allocating (and reallocating) airlift and designing (and redesigning) the movement network. The process centers on integrated operational and combat support planning and incorporates activities for continually monitoring and adjusting performance. A key element of planning and execution is the feedback loop, which determines how well the system is expected to perform (during planning) or is performing (during execution) and warns of potential system failure.

In this chapter, we apply the STT and closed-loop frameworks to theater airlift planning and execution. First, we modify the generic framework to address theater airlift planning and execution-specific objectives and tasks and relate these to higher-level military and national security objectives. Then, we adapt and apply the closed-loop framework to theater airlift planning and execution, focusing on deployment and sustainment. Finally, we extend the STT framework from the task level down to the resource level to link theater airlift planning and execution requirements with the resources available to perform the tasks. In doing so, we apply and modify the framework to deal with the contingency planning and execution time horizon.

The Theater Airlift Planning and Execution STT Framework

Figure 2.1 illustrates a revised STT framework as applied to theater airlift planning and execution.

National Security Objectives

National security objectives are clearly specified in the *National Security Strategy of the United States* (2002, pp. 1–2). This states that the United States will:

- Champion aspirations for human dignity.
- Strengthen alliances to defeat global terrorism and work to prevent attacks against us and our friends.

Figure 2.1
Theater Airlift Planning and Execution Hierarchy of Linkages

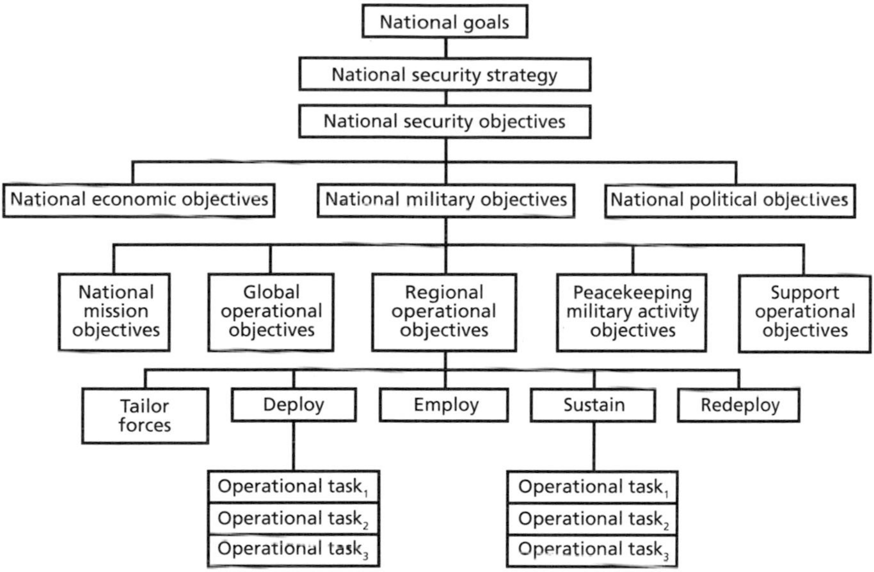

RAND *MG377-2.1*

- Work with others to defuse regional conflicts.
- Prevent our enemies from threatening us, our allies, and our friends with weapons of mass destruction.
- Ignite a new era of global economic growth through free markets and free trade.
- Expand the circle of development by opening societies and building the infrastructure of democracy.
- Develop agendas for cooperative action with other main centers of global power.
- Transform America's national security institutions to meet the challenges and opportunities of the twenty-first century.

National Military Objectives

National military objectives constitute the military component of the strategy to achieve the defined national security objectives. As defined in the *National Military Strategy of the United States* (Myers, 2004), these are to:

- Protect the United States against external attacks and aggression.
- Prevent conflict and surprise attack.
- Prevail against adversaries.

Regional Operational Objectives

The next level in the hierarchy is the operational objectives that define how military objectives will be pursued in the context of a specific theater, scenario, or threat. During Operation Desert Storm, an example of a political objective might be to expel Saddam Hussein from Kuwait. The regional or operational objective was to cut off communications and destroy supply lines. To accomplish the objective, the air component was tasked to maintain air superiority. This was enabled by the specific operations and logistics tasks listed:

- tailor force and support packages
- deployment
- employment
- sustainment
- redeployment.

Tailor Force and Support Packages. Rapidly tailoring force packages requires that the system begin to generate support requirements based on desired operational effects. Combat support planners must coordinate closely with operators to estimate suitable force packages capable of achieving the desired effects while maintaining minimum deployment requirements. Early generation of combat support requirements can contribute substantially to course of action assessment, focusing efforts on feasible courses of action early in the planning process.

Deployment. Rapid deployment requires that combat support planners provide force beddown plans and assessments quickly. Assessments must begin before plans are finalized, and therefore the capabilities and status of all potentially relevant airfields must be available. In addition, the status of in-theater resources must be continuously updated and effectively communicated to facilitate rapid Time-Phased Force and Deployment Data (TPFDD) development.

Employment and Sustainment. Here we define force employment simply as conducting combat operations. We consider sustainment to be all activities necessary to support and maintain employment. This requires that theater and global distribution, maintenance, and supply operations be rapidly configured and expanded and that global prioritization and allocation of combat support resources be rapidly shifted to the area of interest. Effectively allocating scarce resources requires the system to monitor resources in all theaters and prioritize and allocate resources in accordance with global readiness.

Redeployment. As with deployment, this requires that the status of in-theater forces and resources be continuously updated and effectively communicated to facilitate a smooth and rapid transition.

Theater Airlift Planning and Execution Operational Tasks

As we continue down the STT hierarchy, the final level of objectives we consider is theater airlift planning and execution operational tasks. Theater airlift planning and execution operational tasks support the theater airlift planning and execution operational objectives. They are the most specific, and this is the level we break down tasks that require specific combat support capabilities. Our focus here is on deployment and sustainment.

In all, we identify 12 operational tasks, subdivided between deployment and sustainment.

Deployment. ***Allocate Movement Resources***—done by Secretary of Defense and USTRANSCOM, allocate from the global supply of transportation assets (for example, C-17s) that may be devoted to deploying the forces.

Determine FOL Beddown Capabilities for Force Packages—have accessible up-to-date information on ability of FOLs to support different aircraft types and be able to evaluate supporting entire force packages.

Plan Movement Network—based on deployment plans, determine movement nodes, modes of movement, routes, and mobility asset beddown locations.

Facilitate Rapid TPFDD Development—have the methods and tools in place for force tailoring and have systems and institutional processes available that can create actual TPFDDs for use.[4]

Assess Feasibility, Cost, and Time of Deliveries—quickly test feasibility of supporting different operational plans with different networks; the cost in airlift; and estimate performance trade-offs.

Configure Inter- and Intratheater Movement Network—rapidly change transshipment points, mobility beddown locations, hub capacities, and communications infrastructure to accommodate operationally driven changes.

Sustainment. ***Forecast Requirements***—estimate resources needed to support deployed forces.

Receive and Prioritize Demands—with allocated theater airlift resources, determine which movement requests will receive airlift support and determine among those requests what priority they will be given, based on COCOM guidance.

Schedule and Execute Missions—create missions to transport cargo and passengers and determine which aircraft and crews will fly them.

Feedback System Performance—monitor effectiveness and efficiency metrics for feedback to customers and planners.

Assess Network Performance and Reconfigure—assess how well the network configuration of nodes, modes, and routes are serving dynamic customer needs; reconfigure as needed; and request additional resources to do so.

Balance Deployment and Sustainment Resource Allocations—determine impacts of allocating scarce resources between deployment and sustainment to best serve campaign objectives.[5]

---

[4] Snyder and Mills (2004) created a methodology and prototype tool for the Air Force to estimate force packages and their movement requirements from relatively few operational inputs.

Theater Airlift Planning and Execution Force Elements

"Force elements" are the groups of resources (personnel, training, and equipment) needed to perform a task. Because many different types of force elements can be used to support a task, decisionmakers must choose the resource combinations that are most cost-effective and timely in accomplishing a task. Typical theater airlift planning and execution force elements include C-17 aircraft and crews for movement, tanker aircraft to provide an air bridge, C-130 aircraft and crews for tactical transport, and commercial transportation when available.

Theater Airlift Planning and Execution Closed-Loop Planning and Execution Processes

Now that theater airlift planning and execution objectives and tasks have been outlined, we will shift our analysis to a process view by applying a modified closed-loop planning and execution framework (see Figure 2.2) to theater airlift planning and execution requirements and available resources. We first look at deployment and then sustainment.

Deployment

When war plans are formalized, planners choose forces to accomplish campaign objectives, according to the COCOM's campaign plan. Support forces are derived from these operational forces, thereby establishing the total force package. As part of this process, the Secretary of Defense allocates from the entire fleet of mobility assets, in light of global demands, a certain amount of assets to the COCOM planning the war. In network planning, planners determine destinations, intermediate bases, and refueling points. Finally, the TPFDD is generated and executed.

---

[5] In reality, deployment, sustainment, and redeployment all happen concurrently, and much care must be taken in allocating airlift (and other resources) among them.

Figure 2.2
Theater Airlift Planning and Execution Closed-Loop Planning

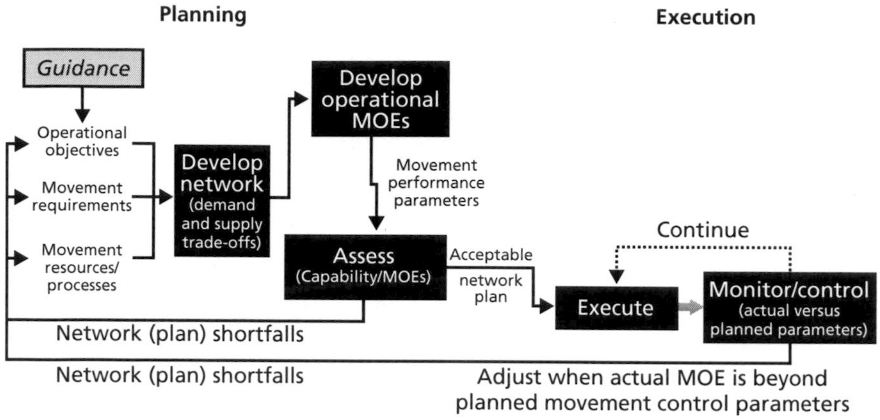

RAND *MG377-2.2*

The COCOM's planners create a means to take deploying forces from APOEs to destinations using sea, air, and land forces. The Secretary of Defense also approves allocation of theater mobility assets (C-130s) for intratheater deployment and sustainment. Theater planners must also devise a deployment network of staging points, airlift aircraft beddown locations, and routes.

Sustainment

For sustainment, we focus on theater airlift planning and execution. Planners start with the forces being supported in theater. They essentially have the network and may adapt it as operations progress. They look at the existing network of FOLs and available airbases. At the tactical level, units (for example, battalions and squadrons) that have deployed and are operating submit cargo and passenger movement requests through their components to the COCOM. The COCOM, through the JMC, establishes movement priorities and determines if a particular request will move by air or not and at what priority. The JMC sends the requests to the CFACC/AMD, which in turn schedules theater airlift missions to fulfill the requests.

As operations progress, demands change. Combat operations shift geographically, and changes occur in operational tempo. As these and other changes occur, the need increases to reconfigure the movement network, including intermediate hubs and beddown locations and sizes, through a closed-loop process. See Appendix C for an example of the closed-loop process applied to movement and support options.

Resource Allocation Within the STT Framework

With the basic theater airlift planning and execution process defined, we now expand the STT framework to highlight the task of resource allocation, which occurs at both the global and theater levels. The basic resource allocation task for theater airlift planning and execution activities can be viewed as a problem of integrating the demand for resources (that is, moving people and cargo, as linked by our framework to higher-level military and national security goals) with supply (that is, processes associated with planning, replanning, and executing airlift operations). Finally, we identify integrator processes (that is, resources for accomplishing tasks). Figure 2.3 provides an illustration of how resource allocation considerations can be integrated into an STT framework that manages contingency movement planning and execution processes.

Demand-side tasks are organized by components to achieve illustrative capabilities (left side of the figure). Force and support elements (right side of the figure) can be selected from component providers to create the capabilities on the left. The integration task is choosing the force and support elements from a list of options, each of which may have differing attributes and offer differing capabilities (middle of the figure). The result of the selection of force and support options, supporting the COCOM's campaign plan, creates the joint operational effects (shown at the bottom of the middle section).

Figure 2.3
Expanded Strategies-to-Tasks Resource Allocation Framework with the Theater Airlift Planning and Execution Framework

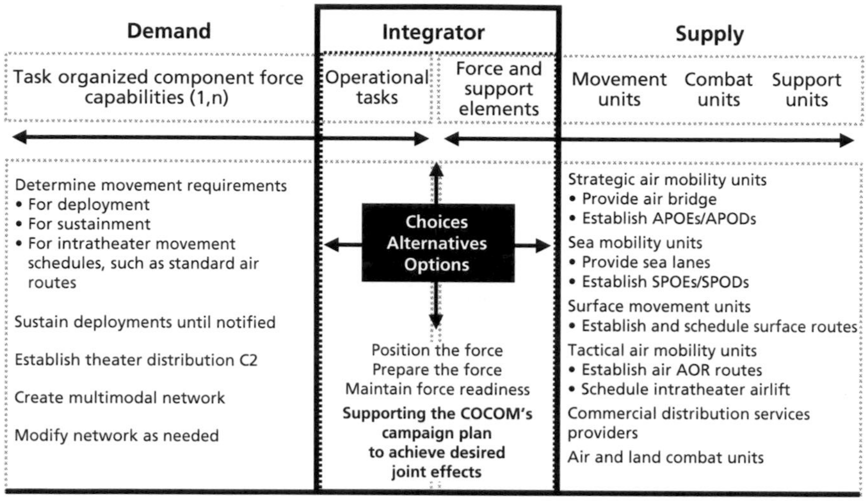

RAND *MG377-2.3*

In this case, movement choices can be made from a set of options including airlift only, sealift only, surface movement only, or some combination of two or three elements. Each choice may result in differing operational task capabilities—for example, different time lines for establishing force presence in theater.

Each of the supply elements is generally provided by a component. Each of the operational tasks, supporting the COCOM's campaign plan, may require combinations of component resources to achieve the desired capability and ultimately the joint operational effect. The conduct of the operational tasks creates movement demands to be supplied, according to the choices made by the integrator.

Figure 2.4 shows a high-level representation of the expanded STT framework that relates movement system supply processes to demand for movement services. We will use this type of representation to present AS-IS and TO-BE process characteristics.

Figure 2.4
Framework for Movement Process Responsibilities

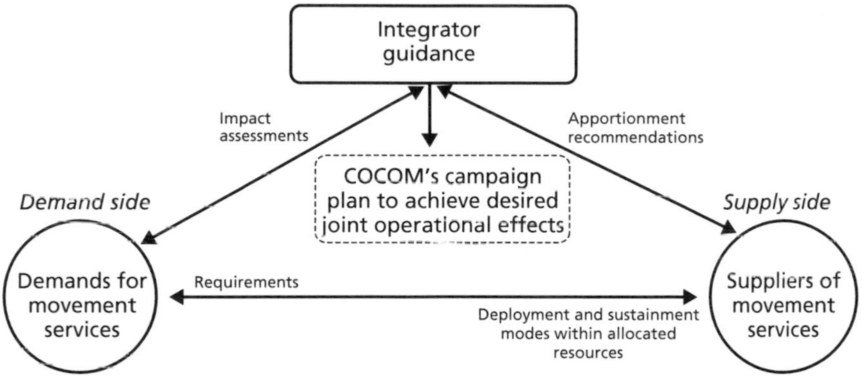

RAND *MG377-2.4*

An important feature of this framework is the analysis and assessment of supply and demand options needed to meet movement requirements. From a strategic level, analyses on the supply side can show how alternative multimodal networks can meet varying movement performance needs, as specified, from the demand side. These alternative networks can affect allocations of modal capacities, transshipment nodes, and routes that can be used to meet movement needs. The integrator chooses the allocations based on analyses of performance and resource needs. Choices must be made that are consistent with allocations guidelines made by the higher authority.

Once the allocations are made, a neutral integrator, who allocates scarce distribution resources among competing demands using the priorities set by the higher authority, prioritizes demands. The integrator needs to be independent from those providing services and those demanding services.

When evaluating how processes should be assigned to organizations (TO-BE) or when evaluating shortfalls in how processes are assigned (AS-IS), we considered two principles in developing a balanced resource STT framework. First, supply-side and demand-side decisionmaking processes should be independent from one another with the integrator making the choices. We call this the "independence principle."

If the integrator is too close to the supply-side processes, decisions may be affected more by efficiency of movement resources and insufficient attention may be given to the effectiveness of choices. If, on the other hand, the integrator is too close to the demand-side processes, effectiveness may be given most attention and efficiency of resource utilization may not receive enough attention.

Second, the demand and supply sides functions should avoid conflating their decisions. The former determines the "what" and "when," the latter the "how." Following this principle, the demand-side processes specify the movement requirements and priorities for movements, including deployment, redeployment, and sustainment needs. The supply-side processes decide how to satisfy those needs. In other words, the demand-side does not tell the supply-side to have 65 C-130s on hand to satisfy movement needs. The supply-side determines the movement vehicles needed to satisfy the movement requirements within the indicated time frame.

With these principles in mind, Figure 2.5 shows a high-level resource-balanced STT framework that relates supply, demand, and integrator processes in the joint military movements arena.

According to our proposed framework, demands are generated by the deployment, sustainment, and redeployment of component task elements assigned to JTFs. The demands for distribution services associated with JTFs are submitted through their component commands to a representative of the COCOM who allocates scarce distribution resources supplied by components on the supply side. The COCOM must live within the movement resources allocated to his AOR by the Secretary of Defense. Supply-side resources are allocated to intra- and intertheater movement capabilities.

Assessment of options for meeting demands is an important feature of this framework. In this case, assessment of supply options should be undertaken to determine how best to meet JTF requirements. From a strategic level, analyses on the supply side can show how alternative multimodal networks can meet varying movement performance needs, as specified from the demand side. These alterna-

Figure 2.5
Supply, Demand, and Integrator Applied to Joint Movement Processes

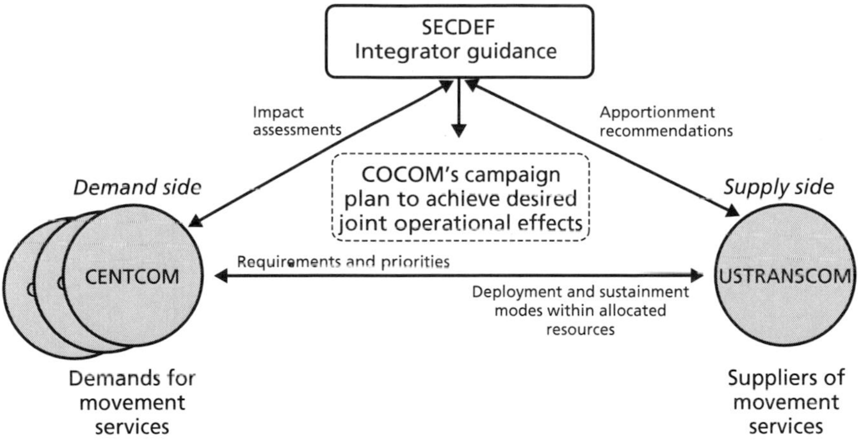

RAND *MG377-2.5*

tive networks can affect allocations of modal capacities, transshipment nodes, and transportation routes that can be used to meet COCOM needs. The COCOM chooses the allocations based on analyses of performance and resource needs. Choices must be made that are consistent with resource allocations guidelines made by the Secretary of Defense.

Once the allocations are made, a neutral integrator, who allocates scarce distribution resources among competing demands using the COCOM set of priorities, prioritizes the demands. The integrator needs to be independent from those providing services and those demanding services, for example, a JMC as described in doctrine.

Nested Responsibilities

Each COCOM, when planning or executing, has movement requirements (aircraft, crews). The Secretary of Defense allocates lift resources among these requirements. If this function is performed correctly, supply and demand are better balanced.[6] This is the start-

[6] Of course, supplies could easily be depleted if demand levels are higher than force structure levels.

ing point for theater airlift planning and execution. The resources allocated to the COCOM are the resources available for airlift operations.

A feature of supply and demand relationships is that they are often nested. In Figure 2.5, we showed USTRANSCOM as a supply-side organization when viewed by the Secretary of Defense or a COCOM. USTRANSCOM does have an integrator role also, as can be seen in Figure 2.6. Each of the DDOCs, for example—CENTCOM DDOC, PACOM DDOC, and so forth—are demand-side organizations when viewed from a USTRANSCOM perspective. In other words, each of the DDOCs is assigned to a COCOM and reports to the COCOM J4.[7] Each of these DDOCs calls on

Figure 2.6
Relationships Are Nested—USTRANSCOM as the Integrator

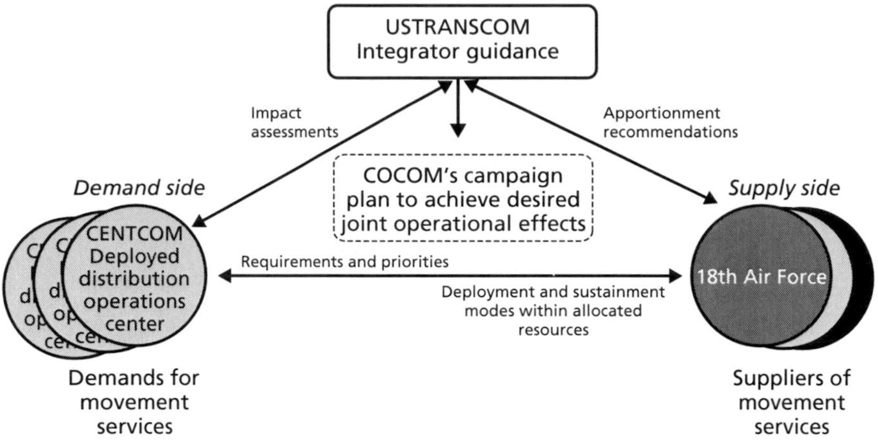

RAND *MG377-2.6*

---

[7] As defined in Air Force Doctrine Document–2, *Organization and Employment for Air and Space Operations*, the Air Force terminology used by the authors identifies organizations/responsibilities. The A/J3 is the Operations Directorate; the A/J4 is the Logistics Directorate; and A/J5 is the Plans Directorate (with A standing for "Air Force" and J standing for "Joint").

USTRANSCOM to secure additional movement resources for their COCOM. USTRANSCOM must balance these needs and assess the impacts of allocating additional resources to one DDOC on the ability of other DDOCs and COCOMs to meet their movement needs.

By the same token, the COCOM is a demander in Figure 2.6. However, from a different perspective, he is an integrator of demands with his own theater resources to meet JTF requirements as shown in Figure 2.7.

The nesting that exists in planning and executing movement adds another layer of complexity to the overall TDS. In Chapter Four, we will discuss some of the AS-IS shortfalls in theater airlift planning and execution.

Heretofore, we have introduced a framework and applied it to the processes associated with theater airlift planning and execution. Next, we identify shortfalls in existing (AS-IS) processes based on our framework.

Figure 2.7
Relationships Are Nested—COCOM as the Integrator

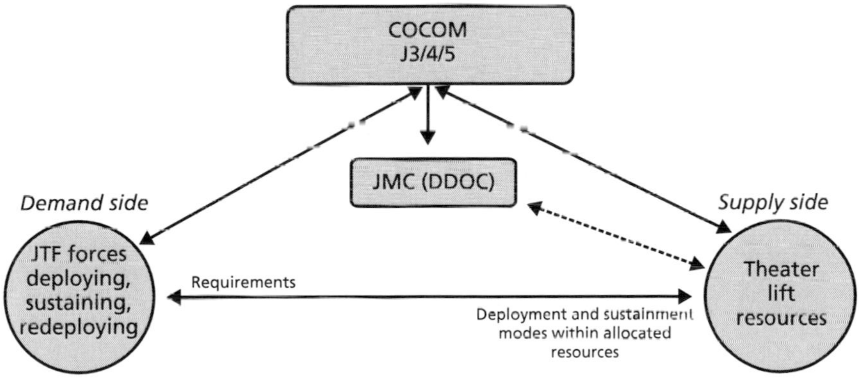

RAND *MG377-2.7*

CHAPTER THREE

Shortfalls in the AS-IS Theater Airlift Planning and Execution System

The objective of our research is to evaluate options for improving CENTCOM's theater airlift planning and execution to support joint expeditionary contingency operations. We now turn our attention to a detailed discussion of the AS-IS theater airlift planning and execution shortfalls[1] and TO-BE suggestions to mitigate these shortfalls. Using the expanded STT framework with the closed-loop planning construct, for each area (supply, demand, and integrator) we will analyze process, organization, doctrine, training, and systems. We will use this framework to evaluate the adequacy of AS-IS distribution planning and execution processes and to identify missing or incomplete processes and deficiencies in doctrine, training, organization, and communications and information systems.

A good deal of progress has been made in addressing the joint distribution process.[2] Establishment of the DDOCs within each COCOM, the move to single destination pallets, and attention to end-to-end performance metrics are concrete examples that clearly demonstrate that improvements in CONUS to AOR distribution processes and performance have been made. Still, numerous shortfalls in the process remain.

[1] When this study began, CENTCOM had theater airlift planning and execution processes in place as outlined in doctrine. As such, the term *AS-IS* refers to both the CENTCOM theater airlift planning and execution processes and the processes outlined in doctrine.

[2] See Appendix F for the evolved CENTCOM intratheater airlift planning processes.

AS-IS Theater Airlift Planning and Execution Process Shortfalls

We begin with AS-IS theater airlift planning and execution process shortfalls. First, demand-side processes are fragmented. Movement requirements and priorities associated with deployments are the responsibility of CENTCOM J3 and the A3, G3, and M3 at the component level. The CENTCOM J4 is responsible for identifying sustainment requirements and priorities. The result of this fragmentation of responsibility is that the integration of requirements and priorities is not always made, leaving unresolved conflicts in movement priorities during execution at the tactical level. Also, the integration takes place on the ramp or loading dock by people without the knowledge of what should go first with limited transportation resources.

Second, supply-side processes are fragmented at the COCOM and component levels. The J4 is responsible for distribution planning, and the J3 is responsible for deployment and redeployment movement planning. At the component level, planning is largely stovepiped by mode—that is, surface transportation, airlift, and sealift.[3]

Third, movement planning and execution processes do not relate how alternative networks or resources dedicated to movement affect the COCOM's campaign plan and ultimately the joint operational effects. In addition, recognition of the need for explicit tradeoffs among deployment, redeployment, and sustainment movements is not apparent. A recent letter from the CENTCOM J4 outlined metrics that should be used to evaluate sustainment movements. The letter specified only effectiveness-related, demand-side metrics—for example, customer wait time and visibility of assets. It made no mention of efficiency metrics and the need to balance movement resource needs with demands for movements. In addition, feedback loops and

---

[3] In EUCOM, J4 planners are embedded in the J3 to help gather and prioritize movement requirements to achieve the operational objectives. This will be discussed further in Chapter Four.

diagnostics have not been established that relate movement performance to the movement parameters needed to achieve the COCOM's campaign plan.[4] This relates to the inability to communicate how movement performance impacts joint operational effects.

Fourth, the use of airlift is not fully integrated into the joint movements system. Intratheater airlift operations and the joint movements system include both distribution and deployment processes. The same movement vehicles satisfy deployment, redeployment, and sustainment movements. Stovepipes exist between the modes, and trade-offs across modes are not fully considered.

Figure 3.1 illustrates some of the major process disconnects in existing, or AS-IS, CENTCOM TDS planning and execution functions when viewed from a multimodal perspective. Text in lighter type indicates assessed shortfalls.[5]

One of the major problems with the AS-IS multimodal TDS planning and execution processes is that current doctrine encourages an ad hoc approach to the structure of theater distribution systems. This doctrine allows the COCOM to select the TDS developer and manager based on which component has the preponderance of force in a given contingency.[6] When applied, this policy can result in different services developing a TDS within the same AOR for different contingencies within a very short time span, as was the case in Operations Enduring Freedom and Iraqi Freedom. For Operation Enduring Freedom, the Air Force was selected to be the component to develop the TDS. For Operation Iraqi Freedom, the Army was selected to develop the TDS. This policy can lead to the use of ad hoc and typically different policies, processes, and capabilities for developing the movement system. Figure 3.1 shows the Army as the TDS

---

[4] Feedback loops relating movement performance to movement parameters did exist in EUCOM in 1999, during operations in the Balkans.

[5] CENTCOM has taken steps to address many of the shortfalls outlined in Figure 3.1. See Appendix F for the evolved CENTCOM intratheater airlift planning process.

[6] Joint doctrine indicates that TDS responsibility can be appointed to any service based on "either the dominant-user or the most-capable-service concept" (Joint Chiefs of Staff, 1996, p. v).

lead component with the responsibility to plan and execute the TDS for the CENTCOM AOR.

The lack of firm policies and guidance that assign responsibilities to a given component or standing joint organization for the development of the TDS could lead to inadequate attention given to this responsibility by all components. Without a permanent champion who has the responsibility for TDS development, inadequate numbers of people with STT and multimodal distribution planning and assessment training and experience can be expected to be available during a contingency. One can also expect that planning and assessment processes will not receive adequate attention. The same is true for information system and tool sets needed to plan and execute TDS responsibilities effectively.

Another high-level problem with the current process is that deployment and redeployment guidance involving the movement of

Figure 3.1
AS-IS View Shows Process Disconnects

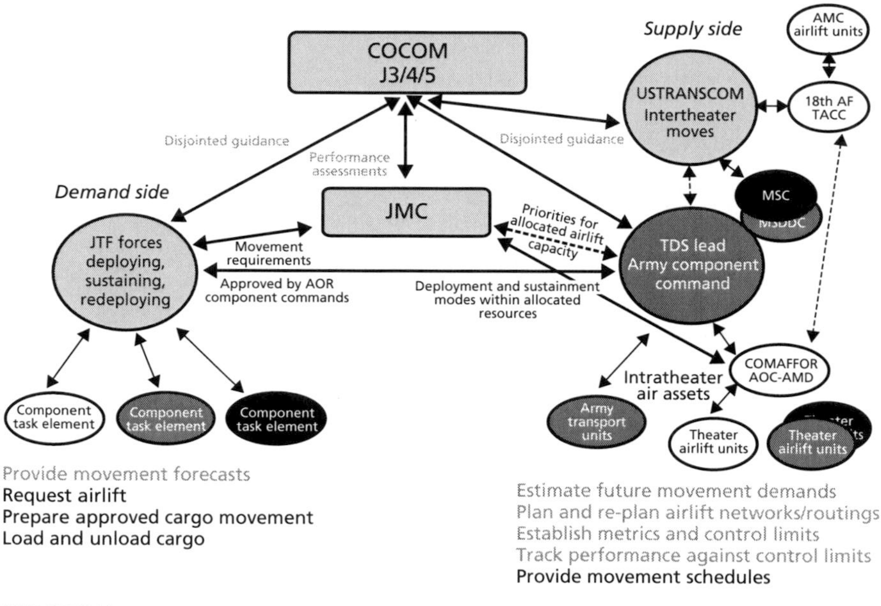

RAND *MG377-3.1*

forces to, from, and within the AOR is provided by the J3. Guidance for the resupply and sustainment of forces comes from the J4. This guidance may conflict and, all too often, may not be resolved by the J3/4/5. As a result, contention for what moves first is resolved by those in the movement terminals, or the first-in, first-out priority scheme may rule.

The left side of Figure 3.1 shows transport demands generated by deploying, deployed, or redeploying force elements. As noted in the lighter type, there are problems in getting realistic and reliable forecasts of movement requirements to develop the intratheater transportation network. Some of these difficulties are associated with the inability to forecast sustainment needs of deployed units. Others are associated with an apparent lack of process, such as an intratheater TPFDD, when a unit moves from one location to another.

The center of Figure 3.1 illustrates deficiencies in integrator processes, namely the ability to relate intratheater transport options to the COCOM's campaign plan and the ability to track performance against what is necessary to achieve desired operational effects.

The right side of the figure shows deficiencies in the supply-side processes for providing intra- and intertheater movement resources. Intra- and intertheater movement system configuration is a complex network design and execution system problem. In military contingency operations, requirements for movement are inherently dynamic because of enemy actions, friendly opportunities that can be exploited, difficulty in predicting demands for such commodities as spare parts, and so forth. The network design must also take into account limited resources and other constraints. For instance, some routings and modal selections might not be available at all times because of threat conditions or time windows for delivery of cargo to forward units. In general, designing military intratheater transportation networks is much more difficult than in commercial activities, where demand patterns are easier to forecast and are much more stable. On the contingency battlefield, forward operating bases can be moved as a result of actions on the battlefield.

As a point of interest, the figure also shows that the DIRMOBFOR (in the AOC) and the AMD do not control all airlift resources that may be available in the AOR, as shown by the dotted lines that have no direct tie to the AMD. For instance, the AMD does not control the use of Army or Navy aircraft or other airlift resources that might be available for use in the AOR.

The lighter type on the bottom right highlights some intra- and intertheater movement planning and execution process shortfalls on the supply side. Most of these shortfalls are related to a lack of a closed-loop, multimodal TDS planning and execution process that can relate multimodal TDS options to the COCOM's campaign plan and TDS performance needed to achieve joint operational effects.

We now use the closed-loop framework to evaluate theater airlift planning and control. Figure 3.2 illustrates some of the key elements of a closed-loop intratheater movement planning and execution process. The text in lighter type highlights missing or incomplete portions in the current planning and execution process.

As shown on the far left of the figure, intratheater movement requirements are hard to predict. Often, these requirements are not provided in detail in the early portions of a contingency operation. In

Figure 3.2
AS-IS Planning and Execution Process Is Not Related to Operational Objectives and Lacks Feedback

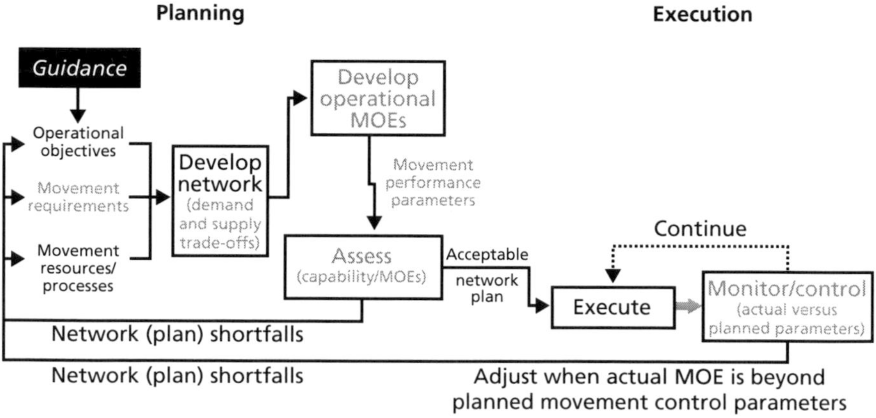

RAND *MG377-3.2*

addition, alternative network designs that adequately show supply and demand trade-offs are not generated.

As shown on the left and center portions of the figure, alternative movement networks are not related to the COCOM's campaign plan to achieve joint operational effects and associated operational measures of effectiveness (MOEs). It is difficult to make these translations, and it will take time to develop a robust capability to do this, but some examples of where the capability exists to relate movement capabilities to combat support effects, such as FOL initial operational capability or weapon system availability, do exist (Amouzegar et al., 2004). We show examples of how this can be done in the next chapter when we discuss the TO-BE planning process. Because movement capabilities have not been related to operational measures of merit, movement control parameters necessary to achieve the desired operational or combat support effects are not specified. Instead, several metrics are collected that are not tied to operational effects, and performance is judged against these metrics. The metrics tend to be demand-side metrics, such as required delivery dates, customer wait times, backlog cargo amounts and time, and visibility of assets in the pipeline. Supply-side metrics are not generally mentioned by the COCOM[7] and neither is the need to balance effectiveness and efficiency. Efficiency is left to the providers to try to attain without high-level attention to the trade-offs.

As shown in the center portion of Figure 3.2, the network planning process should continue until an acceptable plan is attained—that is, one that meets operational objectives with allocated resources. Once an acceptable plan is determined, as shown on the right side of the figure, execution of the plan can be instituted. Data should be collected on the actual performance of the distribution system against the planned control values. When the system deviates from the planned values to a significant degree, operational performance is likely to be affected and replanning actions should take place.

---

[7] See CENTCOM/J4 letter on metrics dated August 2004.

This planning and execution process can be applied at the strategic, operational, and tactical levels of movement planning and execution. More attention is needed at all levels of planning and execution. We provide some examples of strategic- and operational-level planning and execution shortfalls.

While we concentrate on the operational and strategic levels, this is not to say that no challenges in tactical planning and execution exist. For example, determining airlift routings and schedules is a sophisticated mathematical programming problem involving choices among hundreds of variables. Currently, AMD personnel deal with these choices using heuristics and rules of thumb.[8] This may be an area where the application of mathematical programming techniques could offer significant improvements in scheduling effectiveness and efficiency.

Also, improved communication concerning modal performance among destinations could improve intratheater movement performance and result in better decisionmaking on movements. As an illustration, consider the movement of Air Force WRM necessary to establish a primary APOD for Iraq. Establishing this APOD would result in reducing the number of convoys to get cargo to the main Army distribution point.[9] In establishing this APOD, the Air Force A4 requested an early required delivery date (RDD) to move WRM from intratheater storage sites to open the APOD as early as possible. RDD may be a good operationally relevant metric, but all too often these dates are generated in a manner that precludes analysis of movement options to meet the needed delivery date. Thus, intratheater modal choices and airlift choices, in particular, may be driven by RDDs submitted by the demand side. Short RDDs may drive airlift decisions by the components. In our WRM example, the RDD drove the movement to airlift. However, while the movement was approved by the JMC, the movement was given a priority 13 for sus-

---

[8] Conversations with CAOC/AMD personnel.

[9] These sorts of issues may not arise in other theaters, such as EUCOM, where infrastructure is more developed.

tainment moves by air, well down the list on air-eligible cargo. Also, these WRM Cadillac Containers can fit within a C-130, but because of restrictions on C-130s aisle clearance space for crew members carrying sidearms for movement into Iraq,[10] the containers could not be moved by C-130s and had to wait for a C-17 to move them. This complicated the movement of the containers. Because of troop rotations and other commitments, delays in moving these containers were significant. As a result, this outsize cargo waited for an extended period, when it could have moved by surface sooner.[11]

An example of decisionmaking affected by training is in the use of C-17s. C-17s, unlike the C-130, are not assigned to the COCOM, but instead chopped (that is, temporarily allotted) to them for a specific utilization—tactical control. While C-130 capability is a nonrevenue-producing member of the USTRANSCOM Transportation Working Capital Fund (TWCF), the C-17 is required to produce revenue using the following process. Once the TACC receives a validated airlift movement requirement from the DDOC/JMC and the determination has been made this request will be supported by a C-17, the information on the passengers and pallets is loaded by the TACC into Joint Operations Planning and Execution System (JOPES) and assigned a unit line number (ULN). Each ULN has a plan identification designator (PID) associated with it. The PID will determine which branch of service will receive the bill to reimburse the TWCF. During contingencies, much of the billing/accounting is against a Joint Chiefs of Staff project code. These codes are used to help the services identify costs associated specifically with the contingency and to aid in simplifying the reimbursement process. If the PID is associated with a Joint Chiefs of Staff project code, the fees spent should be reimbursed. However, the current process may give the user the idea that use of the C-130s is free (because the costs are

[10] The containers would fit into C-130s, but crew members had to wear sidearms on flights into Iraq, and the container would not allow enough clearance for a crew members to move past the container.

[11] This anecdote supplied by Al Pianalto, record of e-mail transmissions, January 2004–April 2004.

transparent to them) and the use of the C-17 costs the user. These misperceptions can drive the wrong behavior.[12]

AS-IS Process Shortfall Examples

We focus on examples of four major disconnects:

- Attention concentrated on intertheater *distribution* yet intratheater *deployment* dominates movement needs.
- Intratheater airlift demands stated in terms of vehicle requirements rather than desired capabilities hampers airlift planners in efforts to use the right vehicles to meet demands.
- Use of multimodal coordination—switch to commercial air carrier to reduce APOE hold time.
- Impact of network options on the COCOM's campaign plan to achieve the joint operational effect is unknown—airlift beddown impact on operational efficiency.

We will discuss each in detail.

Example One: Deployment Movement Versus Distribution Processes. The first example illustrates how intratheater deployment needs dominate movement requirements, yet after recent operations, distribution processes have received the most Air Force attention. We evaluated an example AEF force package deployed to SWA containing 18 fighters, 3 bombers, 10 tankers, and 2 command and control aircraft. These forces could deploy from the continental United States from such bases as Seymour Johnson AFB, N.C., and Tinker AFB, Okla. In addition, the Army package is a battalion-size combat group, or about a one-third slice of a Stryker Brigade. This force includes 100 combat vehicles, and these can be tailored to meet the specific exercise objective. These forces will deploy from either the United States or Europe.

---

[12] During Operation Desert Storm, each component was allocated a specific amount of the movement requirement (pallet positions). This pallet allocation method may have better satisfied movement needs without having to evaluate every shipment as the components were responsible for determining their own movement prioritization.

The movement of combat support equipment is a significant undertaking in deploying this force package. This amounts to about 4,800 tons of support equipment for Air Force operations and about 5,200 tons of support vehicles and equipment for Army operations.

In fact, combat support resources dominate AEF and Stryker Combat Team deployment movements (see Figure 3.3). Furthermore, most of these moves take place within the AOR, and hence, forward movements dominate deployment requirements.

Although intertheater distribution initiatives received the most Air Force attention, intratheater movement has also received some attention. Intratheater movement can be enhanced by appropriate basing of WRM. RAND Project AIR FORCE analyses of WRM basing options address the needs for supporting routine rotations to deter aggression and for supporting contingency operations. These analyses show that WRM throughput and movement throughput drive joint WRM storage location decisions. We call this the big dis-

Figure 3.3
Intratheater Deployment Requirements Are Larger Than Intertheater Deployment Requirements

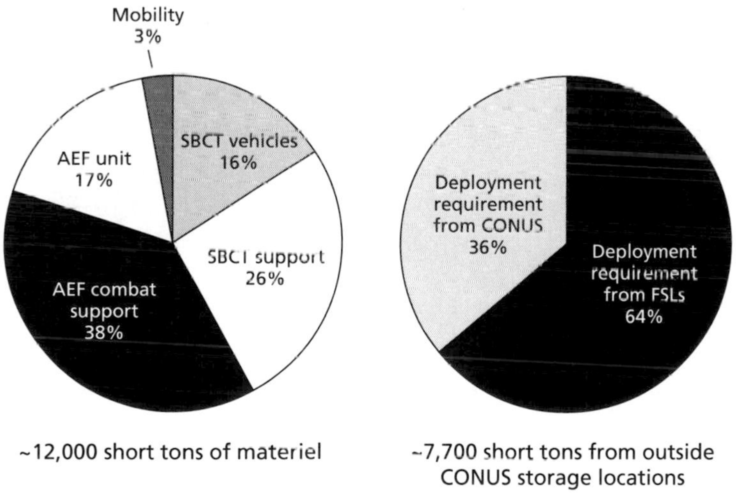

RAND *MG377-3.3*

tribution issue. Yet, attention has been dominated by little distribution decisions concerned with sustainment movement and location of sustainment inventories.

Our work also shows that joint WRM storage locations appear to offer effectiveness and effectiveness benefits, such as reduction in overhead costs and increased throughput by sharing resources, particularly if the sites are at major transshipment hubs with access to land, sea, and air modes of transport. For rapid deployments, some portion of the WRM could be moved by air to support initial operations, while most of the material can be moved by surface means, if threat and political access to land routes and seaports make this feasible.

Example Two: Vehicle Requirements Versus Capability. The second example examines the impact of the demand side stating that a certain number of air vehicles need to be stationed in the AOR to meet intratheater movement requirements. Maintaining a certain number of C-130s in the AOR to handle intratheater airlift needs does not use airlift in the most effective and efficient manner possible. Rather, the demand side should specify priorities and let the CFACC and the DIRMOBFOR (the supply side) determine how to meet the movement requirements—for instance, by scheduling C-17s for movements that may not be resident within the theater.[13]

Taking the deployment example (above) as an illustration, it would take 60 C-130s working for 12 days to finish the WRM movements needed to support the AEF and Stryker force. Alternatively, it would take 12 C-17s to move the WRM within the same 12 days.

The deployment packages to move and support these two airlift options are very different. The aviation and maintenance packages to support a deployment of C-17s are only approximately 200 tons. The aviation and maintenance packages to support a deployment of C-130s are approximately 650 tons. The bare-base assets needed to sup-

---

[13] The Secretary of Defense must approve a deployment order for C-130s, and the TACC has to permit tactical control of C-17s. Both processes require the DIRMOBFOR to work in conjunction with the CFACC, highlighting the need for high-level multimodal planning.

port the C-17 deployment are approximately 2,850 tons, whereas bare-base assets to support a C-130 deployment are approximately 6,340 tons.

The C-130 package is much greater. Thus, the C-17 option would be a much more efficient and effective means of supporting the AEF and Stryker WRM movements. This example shows why the demand side should specify the movement requirements and priorities, not the vehicles that should be stationed within the AOR. The difference is that the COCOM could control the C-130s but would have to rely on the performance of AMC to get the C-17s to the AOR in time to meet the WRM movement requirements in this example. The AMD should have the ability to schedule the vehicles that best meet the requirements—not be restricted to those vehicles that the demand-side specifies should be in the theater—which is the case now in CENTCOM. The Air Force is required to keep 65 C-130s on the ground.

Example Three: Multimodal Coordination. The third example shows how modal decisions need to be made in coordination with regard to the impact on the total pipeline.

In October 2003, the CFLCC requested that a major APOD be opened to serve units deployed to southern and western Iraq. The Corps Distribution Center (CDC) for these units was also moved to this new APOD. The movement of the major Army APOD from Kuwait City International Airport to this new APOD would reduce the need for convoys traveling from Kuwait to the new APOD to resupply units in southern and western Iraq. These convoys were being attacked, and personnel operating the convoys were in danger. Moving the APOD also reduced the risk of having supplies needed by deployed units delayed or destroyed while in convoy.

During this time, surface-to-air missile threats were significant. From October 2003 until March 2004, airlift from Charleston AFB, S.C., directly to the new APOD increased significantly and flights to Kuwait dropped off. Commercial airlift was not used to move cargo into Iraq at that time and still is not used to a great extent because of the threats. As a result, there were significant demands on Charleston's C-17s for these and other movements. Because of other demands, the backlog of cargo destined for the new APOD at Charleston increased in volume and variability. In March 2004, to reduce this portion of the pipeline, APOE hold time, AMC made the decision to use commercial airlift to move a portion of the cargo to the Kuwait City Airport that could not be moved expeditiously by C-17 directly to the new APOD. The TACC coordinated closely with TRANSCOM, CENTCOM Deployed Distribution Operations Center (C-DDOC), and CENTCOM J4 and decided that the use of commercial airlift could be coordinated with the use of empty trucks that were moving between Kuwait and the new APOD.

In this case, the shift to commercial airlift was beneficial because it did reduce Charleston backlog and APOE hold time in the total customer wait time pipeline. This example demonstrates the benefit of coordination between the COCOM and supporting organizations. However, decisions such as this have been made without adequate coordination or analysis of the total affect. Too often, decisions are made on a stovepiped modal basis, failing to take into account the impact on the use of other modes and their capacity and risks. Decisions affecting one segment of the total customer wait time pipeline should not be acted on without taking into account how it would affect the total pipeline. Assessment of operational planning options needs to be strengthened and their impact on total movement performance and ultimately on the COCOM's campaign plan to achieve joint operational effects should be analyzed before decisions are made.

Finally, the issue of who should make movement assessments and choices needs to be addressed. Assessments can be generated by a number of players, but, according to our framework, an integrator who has impact assessments of options and is responsible to the COCOM should make the choices among movement options.

Example Four: Impact of Network Options on the COCOM's Campaign Plan to Achieve Joint Operational Effects. The last example deals with the impacts of being unable to relate airlift beddown options to the COCOM's campaign plan to achieve joint operational effects. Relating movement network options to the COCOM's campaign plan to achieve joint operational effects is important to understanding how movement choices affect the overall military objectives.

An important part of strategic network planning involves selection of APODs, major transshipment points, and aircraft beddown sites. These decisions impact effectiveness and efficiency of airlift operations in the joint multimodal movement system.

In August 2003, approximately 48 C-130s were based out of a major APOD, and another 18 C-130s were based at another location. As shown in Figure 3.4, the second basing location created situations in which aircraft had to fly empty to position the aircraft at APODs or major transshipment points to pick up cargo and passengers for delivery to FOLs in the network. On completing missions for the day, these aircraft returned empty to the basing location. This beddown posture hampered the efficient use of these C-130s.

During this period, the APOD was crowded: F-15s, tankers, strategic airlifters, ISR platforms, and C-130s were based at the air-

Figure 3.4
775th Expeditionary Airlift Squadron Spent Half of Flying Hours Positioning/ Repositioning

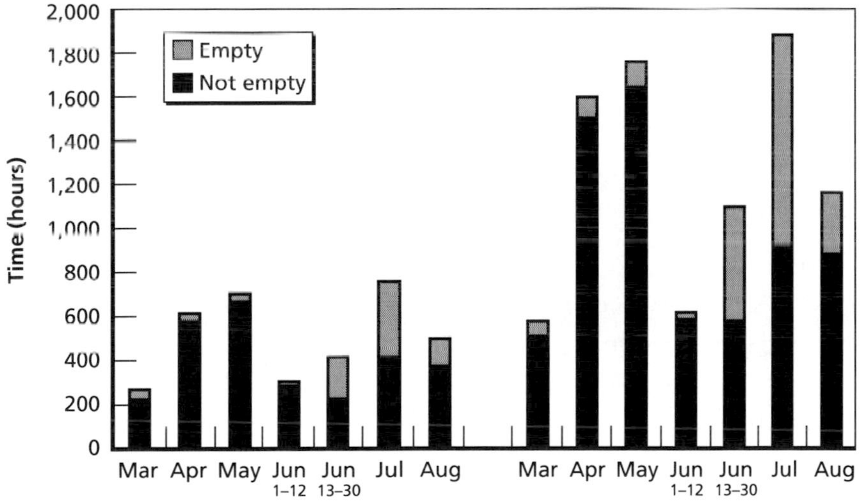

NOTE: Mission considered position/deposition if pallets, tons, cargo, passengers were listed as zero.
SOURCE: MISCAP C-130 Mission spreadsheets.
RAND *MG377-3.4*

field simultaneously. An air base near Kuwait City International Airport supported the beddown of special operations forces (SOF) and Marine aircraft. At the same time, political pressures were being exerted to close bases, and the Air Force wanted to reduce the number of forward-deployed personnel in the AOR. The positioning of C-130s at a major Army APOD and the major distribution site for surface convoys to Iraq may have been a much more effective beddown plan than other basing locations. With a C-130 beddown site at the Army APOD, it is possible that better coordination of cargo between surface and air movement north to Iraq could have taken place and a super-hub operation similar to that in EUCOM between Ramstein and Kaiserslautern could have been developed. After some time, some C-130s were moved to the Army APOD.

The issue here is that intratheater strategy development for airlift does not receive the same attention that strategy development for strike and ISR aircraft do. As a result, airlift advocates do not currently relate airlift options to the COCOM's campaign plan to achieve joint operational effects and thereby do not receive attention for options that affect movement performance. Strike aircraft have limited ranges, limiting their basing options. Airlift aircraft do not have the same limitations. In addition, at the joint level, movement systems are viewed as stovepipes and multimodal planning does not receive enough attention. Not only is this a process void, but it involves joint and Air Force organizational and training shortfalls as well.

In addition, during and after Operation Iraqi Freedom, there was limited understanding of the different types of service that could be provided to common user airlift—for example, frequency-based channels, demand-based channels, and special airlift missions at the component level. The same problems exist at the joint staff level, where a lack of integrating and prioritizing deployment and sustainment priorities can be found. Some users wanted flights every day to their locations and did not have an appreciation of the costs and efficiency of using standard air routes. To improve the utilization of airlift assets, the AMD published a draft Letter of Instruction (LOI) that specified criteria for establishing demand and frequency routes.

Approved by the J4 and the COCOM, the LOI was effective in improving the efficiency of airlift and resulted in the reduction of many scheduled theater airlift routes (STARs) where cargo generation was limited.

The LOI also described the process for obtaining approval for the use of airlift to move high-level cargo, including cutoff times for having requirements delivered to the AMD for scheduling. Processes were also outlined for describing how the components could move cargo ahead of other cargo of the same service that was scheduled to move.[14] This was referred to as the "green-sheeting process." Later, "purple-sheeting" processes were developed that would allow the COCOM to move high-priority cargo ahead of what had already been scheduled for movement by the components.

AS-IS Theater Airlift Planning and Execution Organizational and Doctrine Shortfalls

We now turn our attention to organizations that perform the movement planning and execution processes and the doctrine that guides it. First, movement guidance, including priority guidance, is divided into two organizations. The COCOM J3 is responsible for deployment guidance, and the COCOM J4 is responsible to sustainment movement guidance.

Planning and execution of intratheater movements are accomplished by ad hoc organizations. Joint doctrine gives the COCOM the authority to delegate TDS planning and execution activities to the component with the preponderance of force in the conflict.[15] This promotes an ad hoc approach to the development of the intratheater portion of the joint end-to-end movements system. The

[14] RFID tags and airlift control authority could facilitate movement of higher-priority items.

[15] Joint doctrine indicates that TDS responsibility can be appointed to any service based on "either the dominant-user or the most-capable-service concept" (Joint Chiefs of Staff, 1996, p. v). From 1997 through 2000, EUCOM handled TDS responsibility itself rather than delegating it to a service component.

establishment of X-DDOCs and TRANSCOM's DDOC may be steps in the right direction for making permanent assignment of this responsibility to standing organizations with known missions and staffing, but their functions and responsibilities are not fully developed and no doctrine exists to guide them.

Supply-side, demand-side, and integrator roles of organizations are not clearly understood in terms of the expanded STT resource allocation framework. This framework can assist in thinking through organizational functions.

In addition, strategic thinking about how airlift planning fits into a multimodal joint end-to-end movement system is not fully developed. As a result, how and where airlift planners fit into the joint end-to-end movement strategic, operational, and tactical planning responsibilities have not been fully thought through. These planning shortfalls are particularly acute in strategic- and operational-level planning. As a result, tactical-level planning is left to address problems on a repetitive basis that perhaps could have been avoided if more attention were given to strategic- and operational-level planning.

Finally, reachback is not leveraged to the extent that it could be. Reachback options for airlift planning and execution activity provide an opportunity to accomplish many functions in the rear with the benefit of a reduced deployment footprint and could be more efficient than current ad hoc deployed options.

Limits placed on deploying troop strengths at the highest levels and ad hoc implementation of joint and component responsibilities contributed additional problems in the development of the TDS. The JMC is described in doctrine,[16] but no Joint Manning Document identifies the number and types of people that are needed to accomplish the functions. In addition, the activation is on an ad hoc basis.

Joint Publication 4-01.3 has an extensive discussion of the integrator role of the JMC. It suggests how the JMC Operations Division can be established to prioritize movements consistent with the

[16] Joint Publication 4-01.3 provides detailed allocation process descriptions but very little on network planning and assessment roles.

COCOM's priorities and discusses processes that can be used for this purpose. This publication also indicates that a Plans and Programs Division could be established to facilitate TDS planning and assessment but does not make it a mandatory organization.[17] During Operation Iraqi Freedom, the JMC was activated and staffed on an ad hoc basis with about 50 personnel. Their primary mission was to prioritize airlift movements.

After major combat operations in Iraq, USTRANSCOM was appointed the distribution process owner by the Secretary of Defense. To carry out this responsibility and to assist CENTCOM as much as possible, USTRANSCOM developed the concept of deploying a capability forward to the AOR to help link inter- and intratheater distribution capabilities and resources to better support CENTCOM. The C-DDOC was created and formed with an initial cadre of 60 people to carry out the following mission:

- **Provide** total asset visibility and in-transit visibility of force flow, sustainment, and retrograde.
- **Refine** theater distribution architecture in coordination with the services, Joint Staff, and commanders of JTFs.
- **Synchronize** strategic and operational distribution.
- **Develop** strategic and operational distribution performance measures.
- **Execute** container, 463L pallet system, and Radio Frequency Identification and Detection (RFID) tag management.[18]

The first four mission elements are strategic in nature and are associated with supply-side activities identified earlier. Because the DDOC did not have a clear philosophical underpinning of its role and had to help with JMC staffing shortfalls, it eventually integrated with the JMC and took on its responsibilities. This mixing of supply-

[17] From 1997 through 2001, EUCOM J4 had a Plans Division that operated in this manner.

[18] Taken from C-DDOC Homepage Command Brief (C-DDOC_Homepage_Cmd_Brief_v2.ppt), received from C-DDOC/CC, April 2004.

side and integrator responsibilities may have caused problems with the focus of the DDOC. Much of the time the combined DDOC/JMC dealt with tactical day-to-day decisions concerning the prioritization of airlift cargo. As a result, the strategic shaping of the TDS took a backseat to more immediate concerns of operating the airlift system within the existing TDS.[19]

While the DDOC and JMC were directly involved with the airlift segment of the TDS, development of the surface and sea segments was left to the Army as the lead component for TDS. The practice of dual-hatting several key CFLCC personnel responsible for planning land campaigns with those responsible for planning TDS gave the appearance that the TDS had a decidedly "green" orientation. The separation of the AMD and the other TDS planners may have led to the development of a stovepipe TDS and not an integrated multimodal system with the most effective selection of airlift beddowns, transshipment points, and modal capacities.[20] The planning process deficiencies discussed earlier in this chapter also complicated the development and evolution of the multimodal TDS system.

AS-IS Theater Airlift Planning and Execution Training Shortfalls

The Air Force may need to invest in multimodal training and establish educational identifiers to track training. Currently, the 505th Formal Training Unit at Hurlburt Field, Fla., conducts AOC training. Detachment 1, at the Air Mobility Warfare Center, conducts AMD training. However, training for personnel responsible for airlift planning and execution needs to be enhanced to include methods for integrating airlift into the joint multimodal movement system. Many people dealing with theater airlift planning and execution do not

---

[19] See Appendix F for more discussion on the evolved CENTCOM intratheater airlift planning process.

[20] In Operation Enduring Freedom, the AMD and JMC were collocated. In Operation Iraqi Freedom, they were in different countries.

understand how to apply the expanded resource allocation STT and closed-loop frameworks to maximize efficiency and effectiveness. Classroom instruction on the expanded STT and closed-loop methods and tools needs enhancement. AEF rotation does not allow personnel to become experts. Inexperienced personnel rotate in who are unclear about their responsibilities and ask the same questions with each new rotation.[21] Many people dealing with theater airlift planning and execution could benefit from advanced degree programs that focus on multimodal movement network planning and execution. More training and education is needed on relating movement options and designs to the COCOM's campaign plan to achieve joint operational effects. War games and exercises do not focus on movement requirements and the movement system. Many opportunities arise to address these shortfalls, as discussed in Chapter Five.

AS-IS Theater Airlift Planning and Execution Communications, Systems, and Asset Visibility Shortfalls

This study identified a number of problems with theater communication and visibility. To have communications in theater, both personnel (doctrine, procedure, staffing, training) and equipment (infrastructure and systems) are needed. Conflicts or incompatibilities can lead to degraded quality of communication and visibility.

The process for requesting intratheater airlift is as follows (see Figure 3.5). The unit or user submits a request to the respective A/G4. The request is validated and forwarded to the C-DDOC/JMC. At this point, the request is approved and given a priority. The request and priority are forwarded to AMD for scheduling. Once entered in the ATO, the airlift unit is tasked and the mission is executed.

On the surface, this would appear to be a straightforward operation. However, during wartime operations, the individual services

[21] Interviews with DCFACC, AMC/A45, and AMC/A31.

Figure 3.5
AS-IS Airlift Request Process

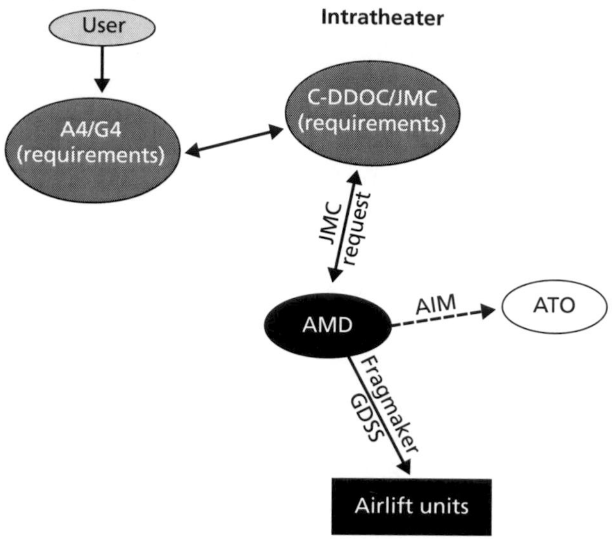

RAND *MG377-3.5*

establish service-specific processes, information systems, and communication capabilities to establish and then transmit airlift requirements, causing this straightforward system to become anything but straightforward.

The way in which C-17 missions are scheduled and controlled further compounds this problem. While the C-130 is the typical airframe assigned to handle intratheater airlift requirements, the C-17, when not performing in a strategic airlift mode, can have nearly triple the carrying capacity of the C-130. C-130s are chopped to the theater and controlled and scheduled by the COCOM's staff. On the other hand, when used for intratheater airlift, C-17s are chopped to the theater, but the TACC controls and schedules them.[22] In this process, the AMD would forward the C-DDOC/JMC-approved airlift request to the TACC for scheduling. Once the TACC has deter-

[22] Currently, the Air Force is experimenting by having the TACC schedule AMC-controlled aircraft.

mined the availability of C-17s to perform the assigned mission, it works with the AMD to get the missions scheduled in the ATO. Typically, this is only done for outsized and oversized cargo or when excessive backlogs begin to build at individual ports.

Figure 3.6 illustrates some of the communications and information system disconnects between the AMD and the component operational units and ATOCs that are operated by the different components. As shown in this figure, different systems and communications architectures are used to carry information on airlift cargo, requirements, eligibility, and status. The Air Force uses Cargo Movement Operations Systems (CMOS) to manage the deployment and redeployment process for its operational units. This system has information on the amount of cargo to be moved and its readiness status. It rides on the unclassified NIPRNET. Global Air Transportation Execution System (GATES) is the sustainment movement system AMC uses to manage sustainment cargo. It also rides on a NIPRNET communications architecture. The Army uses Transportation Coordinator's Automated Information for Movements System (TCAIMS II) to manage its deployment and sustainment cargo, and it rides on a classified SIPRNET communications architecture. The Navy and Marine Corps use Transportation Coordinator's Automated Cargo Information System (TCACIS) and SIPRNET e-mail to manage deployment and sustainment missions. Because these systems operate on different architectures, information gaps exist. These gaps create airlift scheduling and cargo management difficulties.[23]

To add to the system incompatibilities, different services supplied base operating support, which included the communications infrastructure. When each service supplied communications infrastructure to bases, they had their own service priorities, and they supplied the needs of their service-specific information systems. For example, during recent operations in Afghanistan and Iraq, one

[23] JOPES could be used to record intra- and intertheater movements to close shipments. But CENTCOM did not require a TPFDD for intratheater movements.

Figure 3.6
Examples of Information and Communications Disconnects

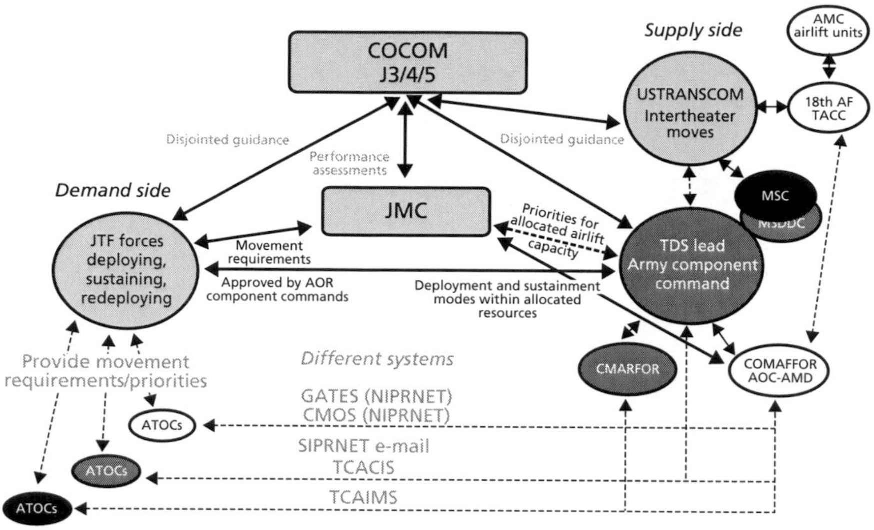

RAND *MG377-3.6*

problem that arose was that when the Army would supply operating support for a base, they often neglected NIPRNET capability, especially early on.[24] Because GATES rides on the NIPRNET, its capability at Army-run bases was often poor, intermittent, or nonexistent.

Other problems also degraded GATES capability. At Kuwait City International Airport, space constraints placed the GATES terminal more than a mile from the flightline. Communications were limited and available only intermittently. These problems had two serious implications. First, visibility was poor at existing locations, a real-time problem for the AMD and others looking for cargo. Further, problems like this (inadequate communications pipelines, intermittent service, poor support, inconvenience) lead to frustration with GATES in general. Complaints mounted about the effectiveness and the availability of GATES. Because of these perceptions, users

---

[24] The Army viewed the NIPRNET capability as an extra, used by personnel for e-mail and other noncritical purposes.

would often simply not use GATES at all, even if it were sometimes available. A simple technology solution to some of these issues would have RFID tags read directly into GATES, eliminating the need for personnel to hand-enter the data. Also, theater planners would sometimes not even ask for GATES service at a newly opening ATOC.[25] Because GATES is not part of the ATOC UTC, commanders had to ask for it specifically. A simple solution would be to add the GATES capability to the ATOC UTC.

Compounding the problem of different information systems is the issue of classification of the data elements. On one end of the spectrum, the USMC does not want to provide the names of passengers manifested for specific flight in an electronic format,[26] and, on the other hand, the CAOC would post the entire next-day flying schedule or fragmentation order (FRAG) on a dot-mil NIPRNET Web page.[27] TCAIMS resides on the SIPRNET and the CENTCOM LOI for movement requests, while not specifically requiring the use of the SIPRNET, references material only available from SIPRNET sites. CMOS resides on the NIPRNET and the Global Transportation Network information is available both in a classified and unclassified mode. These differences in approaches and systems are not just between services but also within the Air Force.

The Air Force Falconer[28] AOC plans call for five nearly identical capabilities to be available globally. While the capabilities may be nearly identical, the manner and systems used to produce them may vary widely. Each theater Air Force staff is developing the processes and tools they need to accomplish the mission, some with little regard for what is being developed in other theaters.[29]

---

[25] Interview with AMC/A3, A4, A5 personnel, April 2004.

[26] Conversation with TACC/XON personnel, November 18, 2004.

[27] Conversations with 908th Airlift Squadron personnel, Maxwell AFB, Ala., November 30, 2004.

[28] A Falconer AOC is attached to the Combat Air Force warfighter headquarters and serves the COMAFFOR. The other type of AOC is a functional AOC such as the TACC, which is part of 18th Air Force and collocated with the AMC staff.

[29] Conversation with TACC/XON personnel, November 18, 2004.

As a result of disconnects during recent operations, ad hoc communications and information forms were developed to transmit unit airlift requests through their component logistics organizations for validation and transmission to the JMC. To alleviate some of these problems during Operations Enduring Freedom and Iraqi Freedom, CENTCOM issued specific guidance on how intratheater airlift request should be processed—including detailed instructions and identification of a standard format for submitting request.[30]

Visibility of cargo and passengers within the theater create additional problems. While the AMD creates the schedules (exactly which people and cargo will move on which mission), ATOCs often improvised. There are channel missions for moving opportune cargo; sometimes missions slip and cargo must be dynamically rescheduled; sometimes cargo does not show up as scheduled[31] and room is left on a plane to move a lower priority.[32] For example, a C-130 aircraft may land at FOL Alpha, expecting to pick up 50 passengers. If those passengers have departed on an earlier aircraft or if they have not arrived from the field, the aerial port director at location Alpha working with the aircraft commander and the loadmaster may decide to move cargo instead. The switch from passengers to cargo will cause the aircrew to reconfigure the aircraft. This process takes time and increases the time the aircraft is on the ground at location Alpha, delaying the aircraft's arrival at location Bravo. During peacetime training operations this is not an issue. However, in a wartime environment with a tightly orchestrated ATO, specific slot times are assigned for takeoff and landing of cargo aircraft to avoid interference with combat sortie generation. Any deviation from the schedule can have a ripple effect on not just the airlift operations but also combat operations.[33]

---

[30] CENTCOM, Intratheater Airlift Letter of Instruction (LOI), October 1, 2003.

[31] Conversations with AMC/A43 staff and Deputy DIRMOBFOR and Director of the AMD.

[32] If RFID tags were read directly into GATES, this would ease the problem for cargo.

[33] Conversations with 908th Airlift Squadron personnel, Maxwell AFB, Ala., November 30, 2004.

While scheduling improvisation helps smooth the flow of traffic, two problems can occur. First, ATOCs exerted control over loading aircraft that went beyond the reach of these improvisations. Different ATOCs were run by different services and groups (for example, the AMC Tanker Airlift Control Element, Air Force transportation squadrons, and Army transportation companies). These groups sometimes followed priorities of their service or base commander that ran counter to AMD instruction.[34] This was a source of frustration for the AMD.

Further, even for valid improvisations, ATOCs would often not inform the AMD of its changes.[35] The AMD could not see all of their missions in available information systems to begin with, so these changes without requisite communication left the AMD even more unaware of what was moving, when, and where.[36] AMD personnel could check some status in GATES and infer movements had occurred, but only half of AMD's missions were updated in GATES. Often the only mechanism to know whether cargo had moved was once the aircraft returned to home station and files a mission report (containing information on pallets and passengers actually moved). Otherwise, the AMD often relied on informal communication to get their feedback—e-mails, telephone calls, and sketchy data with inference.[37]

A good total asset visibility system should include system rules, apportionment of air (number of pallets), and as little en-route handling as possible. Lacking a uniform system to plan and manage airlift operations, visibility of cargo ready for shipment and cargo that was moved became troublesome. For example, some cargo that was reported ready to move could have an aircraft scheduled to pick it up,

---

[34] Interview with CENTCOM/DCFACC and CAOC/AMD personnel, March 2004.

[35] Interview with AMC/A3, A4, A5 personnel, April 2004.

[36] Interview with various AMD personnel.

[37] Interview with various AMD personnel.

but it could have been moved on opportune airlift.[38] In addition, aircraft schedules are included in the GATES NIPRNET, and some concern arose about having flight arrival and departure schedules on NIPRNET, given the security issues at some airfields in Iraq. This raises the issue of whether GATES should ride on the SIPRNET architecture.[39] These issues highlight the absence of an organization with the authority to standardize ATOCs within a theater and could illustrate the need for an aerial port coordination authority.

Not only did different services and groups running the ATOCs sometimes follow different priorities, they were also staffed and trained differently, which can lead to visibility issues. ATOC operations may be staffed with either an Air Force or Army contingent. So, during recent operations, they sometimes had different procedures or assumptions, and they had different expertise. This led to different standards across ATOCs in the theater, which made the AMD's job harder. CENTCOM's overall theater communication and visibility requirement was not given to the units. Finally, component personnel that run the ATOCs are not trained to run an effective and efficient intratheater airlift system as part of the joint end-to-end movement system. The effectiveness and efficiency of each individual ATOC can depend on the experience, branch of service, and leadership of the ATOC director as well as the amount of equipment and number of people assigned to each individual ATOC.[40]

As a result of all these issues, only half of what moved by intratheater airlift was in GATES (see Figure 3.7). Figure 3.7 compares AMD mission spreadsheet data used to schedule the aircraft missions

---

[38] Occasionally, an airlift aircraft will transit an area that has cargo or passengers that need to be moved, but the airlift for these moves has not arrived yet or may not have been scheduled. This cargo can be moved at the discretion of the loadmaster and aircraft commander.

[39] Deployment systems are different for each service and systems were rigged (deployment ULNs were created to move sustainment cargo and vice versa) or ad hoc communications systems developed.

[40] Conversations with 908th Airlift Squadron personnel, Maxwell AFB, November 30, 2004.

Figure 3.7
C-130 Missions in GATES

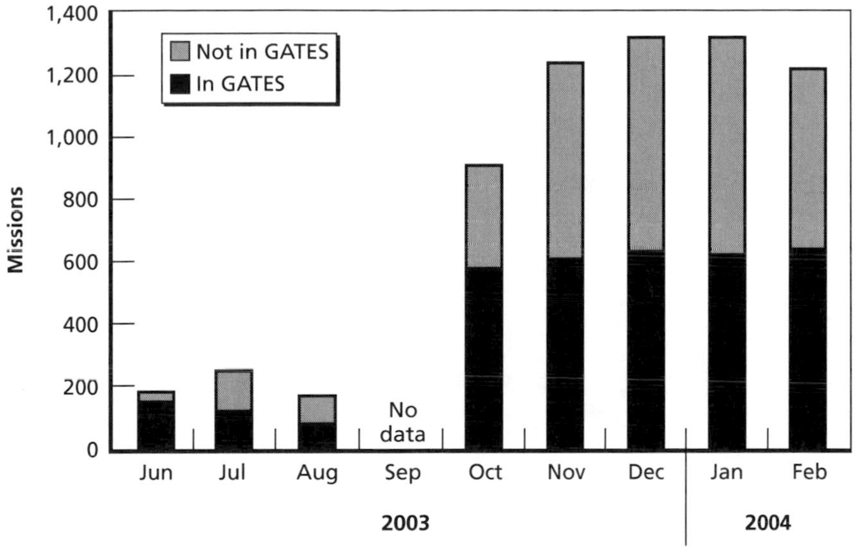

SOURCE: AMD C-130 spreadsheets, RAND GATES database.
RAND *MG377-3.7*

with data that were recorded in GATES over the same period—June 2003 through February 2004. Ad hoc approaches for planning and executing airlift operations were used instead of the system designed to perform this function. This ad hoc approach makes it difficult to plan, replan, and assess airlift operations within the joint end-to-end multimodal movement system.

These communications and systems disconnects make it difficult to determine requirements for airlift and effectively schedule airlift to meet the needs of component operational and support units.

Summary of AS-IS Theater Airlift Planning and Execution Shortfalls

The AS-IS joint multimodal end-to-end movement planning and execution system disconnects are summarized in Figure 3.8.

Figure 3.8
Summary of Theater Airlift Planning and Execution AS-IS Disconnects

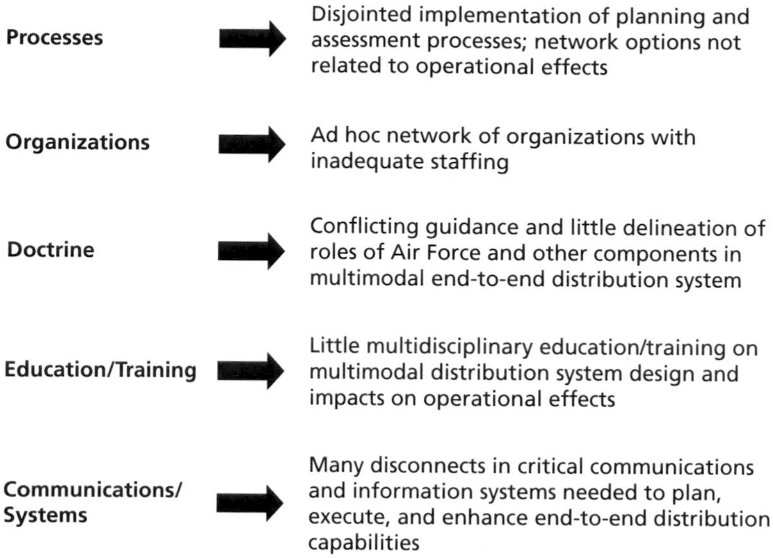
Processes
Disjointed implementation of planning and assessment processes; network options not related to operational effects
Organizations
Ad hoc network of organizations with inadequate staffing
Doctrine
Conflicting guidance and little delineation of roles of Air Force and other components in multimodal end-to-end distribution system
Education/Training
Little multidisciplinary education/training on multimodal distribution system design and impacts on operational effects
Communications/ Systems
Many disconnects in critical communications and information systems needed to plan, execute, and enhance end-to-end distribution capabilities

RAND *MG377-3.8*

In the process area, at the joint level, responsibility for providing guidance is split between the J3 and the J4. The J3 is responsible for providing guidance on deployment and redeployment movements, while the J4 is responsible for providing sustainment movement priorities. As a result, conflicts in guidance are not always resolved at this level, and those providing movement services must contend with conflicting guidance when dealing with constrained resources. Movement planning and execution processes at the strategic, operational, and tactical levels are also incomplete and have not received the attention that combat planning and execution have. Movement network options are not related to the COCOM's campaign plan to achieve joint operational effects. Assessment of options is incomplete. Feedback on planned versus actual performance does not take place.

During recent operations in CENTCOM, the planning and executing of the intratheater portion of the joint end-to-end movement system was ad hoc. Current doctrine supports an ad hoc approach to the development and management of the intratheater movement system by allowing the COCOM to assign the responsibility for TDS to the component with the preponderance of force. The preponderant force shifts among contingencies and potentially within individual contingencies. Thus, no component has the responsibility over time to develop the training, personnel, systems, and tools to accomplish this endeavor. There are guidelines for joint organizations, but they are not mandatory, for example, the JMC has guidelines but no manning document or UTC. Shortfalls have become apparent in the airlift strategy development and coordination of airlift network design options with other modes of movement. Commingling of supply, demand, and neutral integrator processes confuses guidelines about who should be performing various processes.

New organizations have been developed to implement the Distribution Process Owner directive, but doctrine has not been developed for these organizations. Existing doctrine is inadequate on strategic and operational movement planning. Doctrine has not been completed for X-DDOCs and USTRANSCOM DDOC. Doctrine on the role of COCOM in providing integrated deployment and distribution guidance is incomplete.

Currently, not enough airlift planners have training and background on multimodal network planning and execution. More training and education is needed on relating movement options and designs to the operational plan to achieve joint operational effects.

Finally, there are systems and communications disconnects that prevent visibility of movement cargo and passengers. Some of these disconnects result from each service using different information systems to communicate movement requirements and status. These systems ride on different communications architectures, such as, SIPRNET versus NIPRNET. GATES, a primary system used to plan and execute airlift operations, rides on the NIPRNET and many Army combat and support systems ride on SIPRNET. NIPRNET is viewed by the Army as a morale system and is not recognized as an up-front communication system requirement. Spreadsheets and ad hoc procedures were used to develop systems for reporting cargo and passenger status and were not always accurate. All of the above contribute to an incomplete and sometimes inaccurate view of airlift movement.

Process, organizational, doctrine, training, education, communications, information systems, and tools need to be enhanced to improve airlift planning and execution activities within the context of the joint end-to-end multimodal movement system. In the next chapter, we propose options for mitigating shortfalls in the existing theater airlift planning and execution system.

CHAPTER FOUR
Evaluation of TO-BE Improvement Options

In this chapter, we identify options for correcting the deficiencies discussed in Chapter Three. We discuss process, organization, doctrine, training, and system improvements to mitigate the existing shortfalls and evaluate the impacts of the options on joint end-to-end movement system effectiveness and efficiency.

Figure 4.1 summarizes the results of our analysis and indicates that several actions can be taken to enhance airlift planning and execution.

Process, organizational, doctrine, training, education, communications, information systems, and tools need to be enhanced to improve airlift planning and execution activities within the context of the joint end-to-end multimodal movement system.

TO-BE Process Improvements

The TO-BE closed-loop joint movement contingency planning and execution process is illustrated in Figure 4.2. The TO-BE closed-loop process addresses the shortfalls that were discussed in the AS-IS process (detailed in Chapter Three).

As shown on the far left of the figure, integrated intratheater movement requirements need to be provided to end-to-end multimodal network planners. The requirements need to include estimates of time-phased sustainment as well as deployment requirements. This

Figure 4.1
Options for Improving the Joint Multimodal Movement System

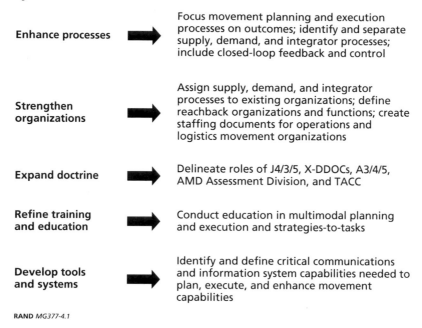

Enhance processes	Focus movement planning and execution processes on outcomes; identify and separate supply, demand, and integrator processes; include closed-loop feedback and control
Strengthen organizations	Assign supply, demand, and integrator processes to existing organizations; define reachback organizations and functions; create staffing documents for operations and logistics movement organizations
Expand doctrine	Delineate roles of J4/3/5, X-DDOCs, A3/4/5, AMD Assessment Division, and TACC
Refine training and education	Conduct education in multimodal planning and execution and strategies-to-tasks
Develop tools and systems	Identify and define critical communications and information system capabilities needed to plan, execute, and enhance movement capabilities

RAND *MG377-4.1*

Figure 4.2
The TO-BE Process Integrates Assessments into Plan Development and Includes Feedback Loops

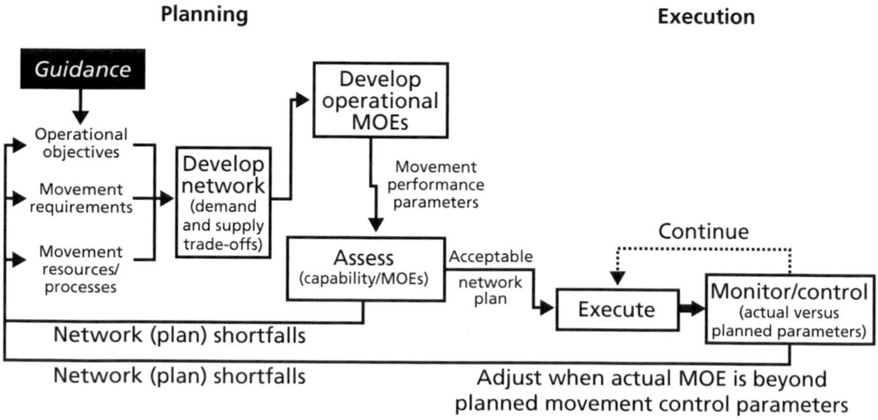

RAND *MG377-4.2*

integrated set of requirements can be provided by the COCOM J3. In addition, alternative network designs can be generated by joint strategic multimodal planners that show supply and demand tradeoffs. We will discusses options that address what organization could conduct these enhanced joint planning and assessment tasks later in this chapter.

As shown on the left and center portions of Figure 4.2, the future process should relate alternative movement networks to the COCOM's campaign plan to achieve joint operational effects and associated operational MOEs. To develop these relationships fully, education programs and time will be needed, but currently available supply and demand related metrics can be used to associate how movement network options are likely to impact operational objectives. For instance, such metrics as end-to-end time for movements, which are functions of movement options, can be related to FOL initial operational capability and weapon system availability.

As shown in the center portion of the figure, the network planning process should continue until an acceptable plan is attained—that is, one that meets operational objectives with allocated resources. Once an acceptable plan is determined, as shown on the right side of the figure, execution of the plan can be instituted. Data should be collected on the actual performance of the movement system against the planned control values. When the system deviates from the planned values to a significant degree, operational performance is likely to be affected and replanning actions should take place.

This planning and execution process needs to be applied at the strategic, operational, and tactical levels of movement planning and execution. In this section, we provide an example of operational- and tactical-level planning and execution from a component Director of Logistics viewpoint, once strategic movement network decisions have been made. While we present an example on TO-BE operational and tactical planning and execution, more attention is needed on decisions that should be made at each level of planning and execution—strategic, operational, and tactical.

In addition, both supply- and demand-side metrics are needed to assess the effectiveness and efficiency of joint multimodal end-to-end movement system options. Some of the important supply- and demand-side metrics are:

- supply-side metrics
  - airlift effectiveness and efficiency
  - APOE/APOD throughput
  - backlog cargo
  - crew utilization
- demand-side metrics
  - on-time deliveries
  - time required for various priority groups
  - variability in delivery times
  - "operational effects" of transport performance.

To make rational resource allocations, not simply allocations based on subjective judgments or skills of advocacy, decisionmakers should have some understanding of the relationships between movement resources and higher-level operationally related metrics. Figure 4.3 brings together demand-side and supply-side metrics and shows the notional trade-offs between resource allocations made to movement and how they could affect the amount of inventory needed in the AOR to support combat operations.

Figure 4.3
Complementary Resource and Operational Movement Metrics Enable Rational Resource Trade-Offs and Allocations

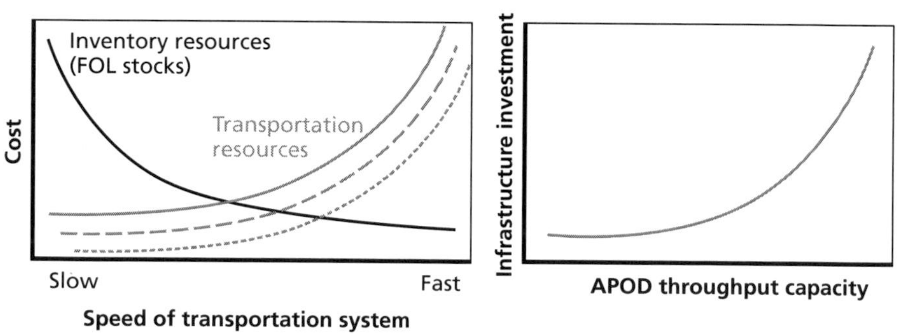

RAND *MG377-4.3*

The graph on the left shows inventory investment in the AOR and how they may be affected by resupply time as a function of multimodal infrastructure investment—that is, the number of spare engines, other spare parts, munitions, and so forth is reduced as resupply time is decreased. Fast resupply is a function of multimodal transportation resources allocated to the AOR—for example, the number of C-130s or surface vehicles allocated for use in the contingency or infrastructure investments in the APOEs/APODs.

The graph on the right shows an illustration of how investments in intratheater airlift infrastructure might affect APOD throughput. Each point on this curve will result in a transportation resource curve on the left graph. Here we show three such curves. The point is that there is no right or wrong answer, but rather trade-offs exist and they need to be considered when making network design option decisions.

TO-BE Process Improvement Challenges

Challenges must be addressed to achieve TO-BE process improvements. First, integrated and improved deployment and sustainment movement requirements are needed. As we discuss in the next section ("TO-BE Organizational Improvement Options"), the COCOM J3 could be made responsible for providing integrated and improved forecasts of movement requirements to those responsible for joint multimodal end-to-end movement planning and execution activities.

Next, training on how movement network options and performance can be related to operational effects is needed. Without adequate training, personnel will be unable to make the important trade-off decisions necessary during contingency operations with limited movement resources. In addition, more people need training and education on how to plan and execute multimodal transportation systems.

Movement metrics need to show trade-offs in performance and resource requirements. There is no right answer on movement speed or capacity. Trade-offs must be made on movement speed and operational effects, costs, and impacts on other AORs of allocation of scarce resources. Understanding the need to make trade-offs should be embraced at the highest levels, and assessments of options should be routine.

Data are necessary for planning, execution, and control of the multimodal movement system. Plans for supplying communications and information systems for supplying these data in the initial phases of contingency operations are needed.

Finally, good linkages and interfaces are needed between those planning combat operations and those responsible for planning and replanning the movement network to support operations in dynamic contingency environments. This will require good organizational, information, and interpersonal interfaces.

TO-BE Organizational Improvement Options

We now turn our attention to organizations that will perform the movement planning and execution processes. To address AS-IS organizational shortfalls, we consider the following options:

- Assign planning processes to existing COCOM staff organizations and improve interfaces as the expanded resource allocation STT framework suggests with execution being conducted by the components.
- Assign planning processes to a line organization that would report directly to the COCOM and be responsible for end-to-end movement planning with execution being conducted by the components.

Both options use the expanded STT framework as a guide and assign movement planning and assessment responsibilities to the COCOM and staff and execution process responsibilities to joint and component organizations consistent with the demand, supply, and integrator roles.

Planning Responsibilities Assigned to Existing Organizations Following the Expanded STT Framework

The first option clarifies movement planning and execution processes and modifies process assignments among existing organizations. Figure 4.4 shows a high-level framework for improved interfaces among supply, demand, and integrator functions.

At the theater level, the COCOM's staff is the integrator providing deployment and sustainment movement guidance. We place the responsibility for estimating integrated deployment and sustainment movement requirements in a new J3 organization, shown on the demand side. This organization, which we have called the Requirements Integration Organization, is responsible for working with JTF commanders to forecast and integrate current deployment, sustainment, and redeployment movement requirements. The J5 is responsible for forecasting and integrating future campaign deploy-

Figure 4.4
Assigning Processes to Existing Organizations Using the Expanded STT Framework Streamlines Joint Movement Responsibilities

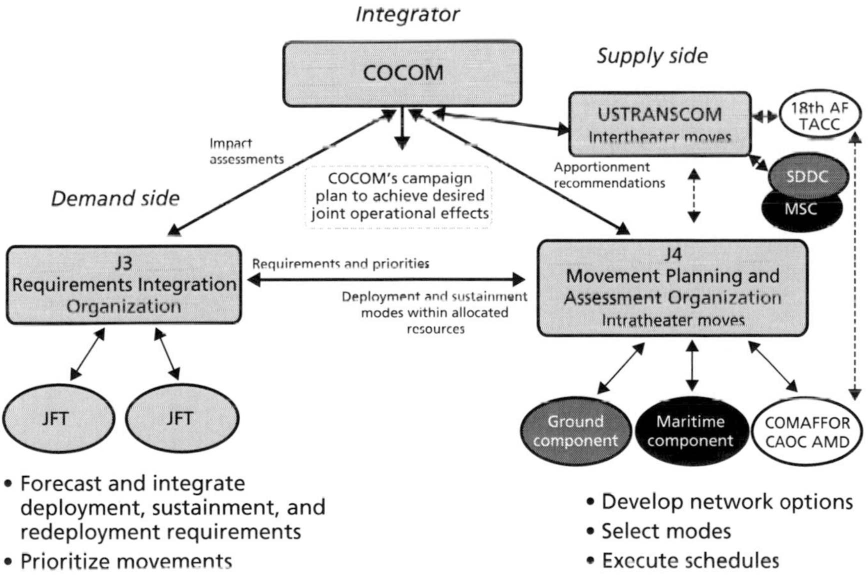

RAND *MG377-4.4*

ment, sustainment, and redeployment movement requirements. Both the J3 and the J5 are also responsible for providing integrated priorities for movements to the J4 Movement Planning and Execution Organization and USTRANSCOM. Embedding a group of J4 planners within the J3 organization could allow J3 planners to focus on operations while the embedded J4 planners focus on prioritizing movement requirements.[1] The J3 Requirements Integration Organization remains a staff-level function in this option.

USTRANSCOM and the J4 Movement Planning and Execution Organization are supply-side organizations responsible for developing movement network options, setting modal capacities, establishing major transshipment points, and overseeing the execution of routes and schedules. In this framework, USTRANSCOM is responsible for planning and assessing intertheater movements and the J4 organization is responsible for planning and assessing intratheater movements. These two organizations must work closely with each other to develop integrated inter- and intratheater movement networks, system options, and plans to execute dynamic contingency requirements. The J4 Movement Planning and Execution Organization would remain a staff-level function and the component commands would remain responsible for executing the plans developed by these joint organizations in this option.

Planning Responsibilities Assigned to a New End-to-End Line Organization Following the Expanded STT Framework

As in the first option, we use the expanded STT framework to separate supply and demand processes and assign these processes to a new organization following the expanded STT framework. We call the new organization the Deployment and Distribution Movement Organization (see Figure 4.5). This new organization is a line organization headed by a Director of Deployment and Distribution Movement. The director position could be a general officer working

[1] J4 planners were embedded in the J3 during operations in the Balkans in 1999 (Discussions with former EUCOM Deputy J4, March 2005).

Figure 4.5
Creating a New Line Organization for Movement Planning and Allocation

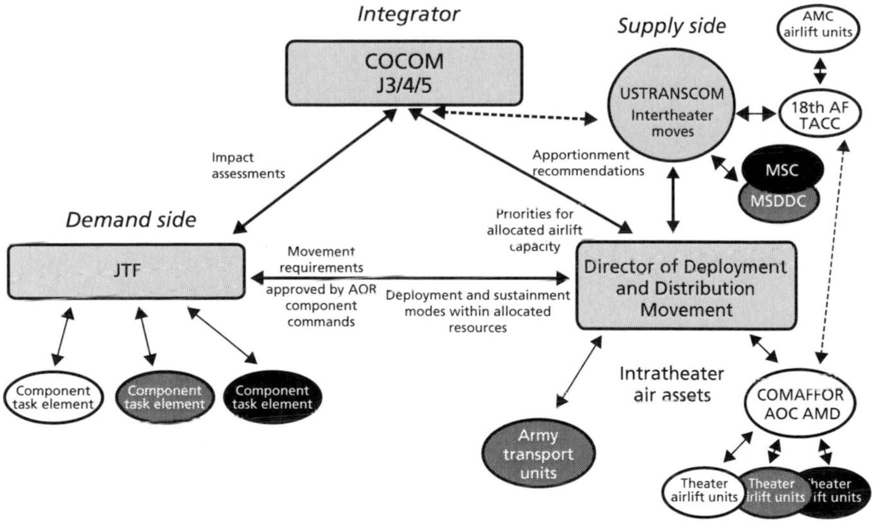

RAND *MG377-4.5*

directly for the COCOM to ensure that COCOM priorities were met, but the individual would also report to USTRANSCOM (the director would be dual-hatted). The Director of Deployment and Distribution Movement would be responsible for planning and assessing end-to-end deployment and distribution movement processes.

The Deployment and Distribution Movement Organization would be responsible for developing movement network options and planning and assessing theater movement—previously assigned to the J4. This organization would also take responsibility for forecasting and integrating current deployment, sustainment, and redeployment movement requirements—previously assigned to the J3. The J5 would still be responsible for forecasting future campaign movement requirements, and the J4 would still be responsible for all other logistical responsibilities except for movement planning and assessment. The components are responsible for execution, as currently specified in doctrine. Figure 4.6 outlines the responsibilities of the Deployment and Distribution Movement Organization.

Figure 4.6
Deployment and Distribution Organization Construct

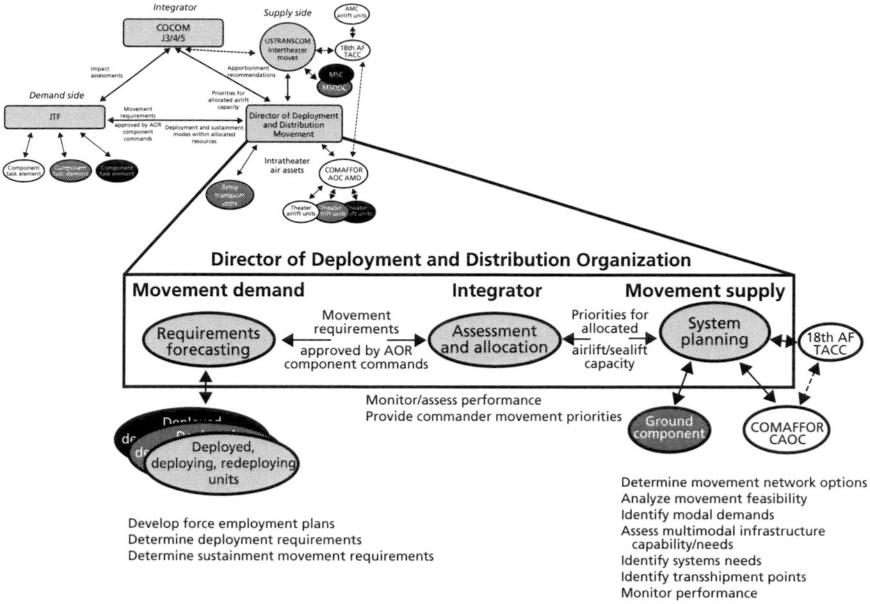

RAND *MG377-4.6*

The responsibilities of what we call the Assessment and Allocation Organization[2] in this option are somewhat clearly identified in Joint Publications as those associated with the Operations Division of the JMC.[3] This organization would assess impacts of system configurations on demand satisfaction and utilization of resources, provide recommendations on system design, and prioritize movements for COCOM.

The responsibilities of what we call the System Planning Organization are not as clearly identified in Joint Publications, but could

---

[2] We use different names, of our own creation, to describe these organizations to avoid comparisons of what was intended for X-DDOCs and our proposals. We believe the names that we have chosen are descriptive of the process responsibilities of these organizations.

[3] Joint Publication 4-01.3 describes these functions as responsibilities of the JMC Operations Division.

be read into the features of the Plans and Programs Division of the JMC[4] which include

- multimodal intratheater movement planning and options development, including resource utilization assessments;
- personnel from each component with needed experience and training—multimodal distribution systems planning and STT capabilities; and
- the proper tools and information capabilities.

These system-planning responsibilities are clearly among those intended for the X-DDOCs as outlined in initial mission statements for the CENTCOM and other DDOCs.[5]

The Director of Deployment and Distribution Movement and staff could integrate and prioritize movement demands and develop movement options and associated resources necessary to meet COCOM requirements. During contingency operations, the general officer in this position could move forward, if necessary, with the COCOM. Being dual-hatted with USTRANSCOM, the Director of Deployment and Distribution Movement could also integrate intertheater movement requirements from USTRANSCOM, making this organization responsible for end-to-end movement processes.[6] This option allows movement requirements representation at a higher level within the chain of command. This concept is being implemented, in part, by the new DDOC organizations in each AOR.

---

[4] Joint Publication 4-01.3 outlines an optional JMC Plans and Programs Division that could include these functions.

[5] Joint Publication 4-01.3 outlines an optional JMC Plans and Programs Division that could include these functions (C DDOC_Homepage_Cmd_Brief_v2.ppt, received from Brig Gen John C. Levasseur, C-DDOC/CC, April 2004).

[6] The dual-hatted Director of Deployment and Distribution Movement Organization should facilitate end-to-end movement planning and execution resource utilization while maintaining a strong COCOM influence on resource allocations among COCOMs.

Another Organizational Option

The first two options outlined above use the enhanced STT framework as a guide to assign demand, supply, and integrator roles among existing organizations. Execution in both options is left at the component level. We now briefly discuss another organizational option: creating a Joint Theater Logistics Commander (JTLC).[7]

Creating a Joint Theater Logistics Commander could create an additional layer of movement command and control (see Figure 4.7). Because the JTLC would be responsible for much more than just

Figure 4.7
The Movement Planning and Assessment Organization Facilitates Development and Assessment of Options

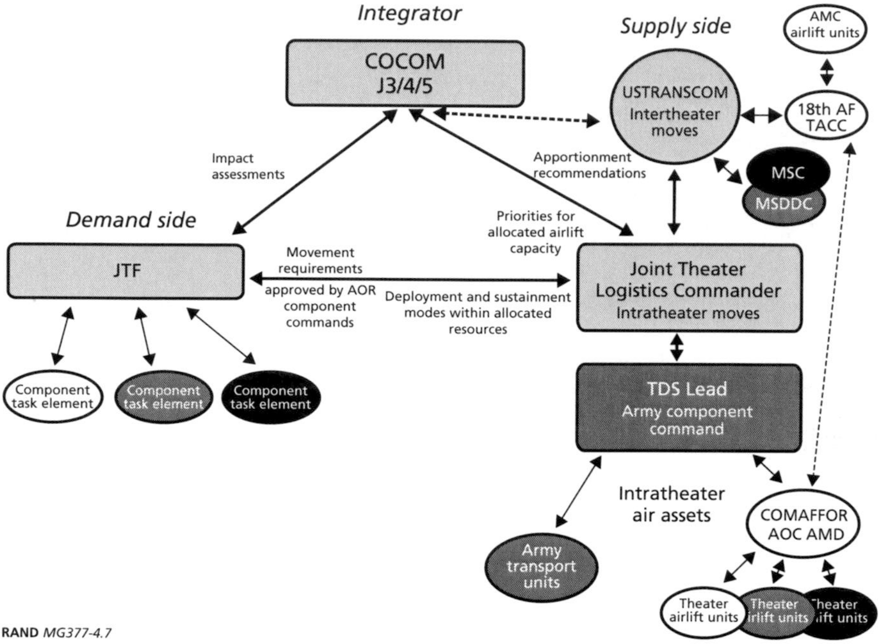

RAND *MG377-4.7*

---

[7] Because this report focuses on movement, we will only discuss aspects of a JTLC related to movement functions and processes.

movement—for example, contracting, base operations, and so forth—the JTLC would likely designate a TDS or movement lead. That TDS lead would then be responsible for movement functions and processes, reporting to the JTLC. In addition, execution would be removed from the components. Assets would be chopped to the JTLC for allocation and execution.

At best, the JTLC would replace our proposed Deployment and Distribution Organization. At worst, it would create additional management layers and communications requirements. Even if this construct were implemented, to be successful, it would need to adopt the improvements suggested by the expanded resource allocation STT framework. It may therefore, be prudent to implement the expanded STT framework before considering developing a JTLC.

Applying Reachback to the STT-Suggested Improvements

In each of these options, opportunities exist for exploiting reachback. Using reachback to CONUS for some of the movement planning and execution functions now performed forward in the AOR requires the sensible division of functions forward and rear. Reachback also raises questions about standard organizational approaches and processes across AORs for providing reachback services.

Reachback can be used to facilitate movement planning and execution processes. For example, the J4 Movement System planners could reach back for planning and assessment support. The J4 planning could establish tentative transshipment points and forecasts of demands for each mode and then have the detailed modal plans take place by the component responsible for supplying the mode, in this case the COMAFFOR. This reachback planning would be iterative in nature and would iterate until an acceptable joint movement plan was developed. This scheme could result in more efficient use of scarce planning talent than the dedication of the staff to the J4 System Planning Organization.

Recently, the CENTAF AMD established reachback support with the TACC. The TACC is currently supporting the AMD by processing diplomatic clearances and planning and scheduling C-17s assigned to move intratheater cargo and passengers in SWA using the C-DDOC (JMC) priorities.

Many other AMD products and services might be supplied through reachback to reduce forward footprint. Reachback could be extended to C-130 and tanker scheduling. However, TACC personnel would need to move away from three-day advance-notice, static planning to a more dynamic one-day planning. Training and education would aid in making this shift.

Care must be taken to ensure that personnel performing reachback services and supplying products know for whom they work. These reachback personnel support the COMAFFOR, not the 18th Air Force commander, not the TACC commander. They are COMAFFOR support personnel.

When determining reachback products and services, care must be taken to ensure that key mobility personnel needed for interacting with the CFACC are deployed forward with the CFACC. Thus, reachback is not an all-or-nothing proposition. Rather, it involves determining the products and services that do not require face-to-face interaction.

Reachback is now underutilized. Many more functions could be supported through reachback. Supplying reachback products and services can significantly reduce AMD's forward deployment footprint and offer economies of scale to provide these services with fewer people. These reductions in personnel required for AMD operations could be converted to supply additional airlift planners needed in the A3/5, J4, and USTRANSCOM positions. In Appendix D, we provide initial estimates of how reachback can reduce AMD personnel requirements and how those positions could be used to enhance COMAFFOR and joint airlift planning within the end-to-end joint movements system.

Assessment of TO-BE Organizational Improvement Options

Based on our analysis, using an expanded resource allocation STT framework to separate supply, demand, and integrator processes and assign them to existing organization to improve effectiveness and efficiency should be implemented. Just applying the expanded STT framework should help the J4 do his or her job. However, creating a Deployment and Distribution Movement Organization will also work. With either option, a thorough review of possible reachback options should be completed. Implementing the expanded STT framework, using either organizational option, brings several advantages.

First, assigning processes to existing organizations and improving interfaces remains consistent with time-tested doctrine that has guided contingency operations for many decades. Basic doctrine calls for COCOMs to develop and execute contingency plans subject to oversight by the Secretary of Defense. Using this doctrine, the COCOMs are responsible for employing forces. The components are responsible for providing forces. Using our expanded STT framework, the COCOM is a demand organization. The components and the unified and functional joint commands, such as USTRANSCOM, are supply-side organizations. Thus, at the highest level the Secretary of Defense is the integrator among COCOM demands and component and specified joint command suppliers.

Applying the expanded STT framework to modify process assignments to existing organizations also insures that COCOM priorities are met by assigning intratheater resources under the operational control of the COCOM and having access to agreed-on or arbitrated allocations of intertheater movement resources. Given adequate planning and guidance, these improved interfaces will support agility in meeting dynamically changing battlefield conditions by having in-theater movement resources under the control of the COCOM.

Realigning processes also strengthens joint strategic and operational level planning and assessment while leaving tactical planning and execution responsibilities in the hands of the components, preserving unity of command.

Applying the expanded STT framework is also relatively easy because it deals with changing processes and clear assignment of these process responsibilities to existing organizations. It would assign intra- and intertheater movement planning and execution responsibilities to standing organizations in each COCOM, at USTRANSCOM, and within each component.

This change also has significant training implications for each of the components, and the Air Force in particular—to provide each COMAFFOR, COCOM, USTRANSCOM, and the 18th Air Force with trained personnel who are educated on multimodal movement planning and execution and STT methods and tools. Communications and information system connectivity also needs to be enhanced.

We have conducted assessments for implementing the expanded STT framework and find that enough people are probably involved with joint movement planning but they are not working the system as the framework would suggest.

We have also investigated the possibility of conducting more airlift planning and execution processes in the CONUS and providing products forward to the CFACC, DIRMOBFOR, and the director of the AMD. We found that reachback can result in more effective use of resources, lower the forward footprint, and offer the potential for standardizing AMD processes worldwide.

TO-BE Organizational Improvement Challenges

To improve airlift planning and execution within the joint multimodal end-to-end movement system, the following actions are needed to modify process assignments within existing organizations.

- Enhance airlift planning expertise within the COMAFFOR A3/5. Some additional people may be necessary to accomplish these functions.
- Create assessment capabilities in the AMD. An Assessment Cell should be created and staffed with a small analysis team.
- Separate supply-side network planning responsibilities, J4 System Planning (currently in X-DDOCs), from assessment and allocation responsibilities (JMC's responsibility, as outlined in Joint Publications). This move does not affect staffing requirements, but the J4 System Planning Organization will require some of the best-educated and most highly trained airlift planners. Some of these planning functions could be supported

through reachback to the COMAFFOR A3/5 enhanced staff and to the TACC and USTRANSCOM DDOC.

- Create a J3 organization to perform integrated requirements forecasts and guidance (demand-side).
- Establish the J4 as the integrated COCOM movements planning and execution supply-side focal point.
- Create reachback communications and define reachback responsibilities and organizations.
- Staff supply, demand, and integrator organizations with people trained in STT, assessment, and multimodal planning.

We address the initial cost estimates associated with these enhancements in Appendix D of this report.

TO-BE Doctrine and Training Improvement Options

Current doctrine allows the COCOM to select the TDS developer and manager based on which component has the "preponderance of force" in a contingency,[8] which can lead to the use of ad hoc policies, processes, and capabilities for developing the TDS. To effectively implement either the improved interfaces option or the creation of the Deployment and Distribution Movement Organization, doctrine must be revised to address the shifts in supply, demand, and integrator processes. Doctrine and Joint Publications need to be refined to address the roles of the J3 Movement Requirements Organization, the J4 System Planning Organization, USTRANSCOM DDOC, and the components. The improved interface option is consistent with current doctrine, but the doctrine needs to be updated to reflect the responsibilities of these current organizations and practices. Joint Publications, such as 3-0, *Doctrine for Joint Operations*; 4-0, *Doctrine for Logistic Support of Joint Operations*; and 4-01.3, *Joint Tactics,*

[8] Joint doctrine indicates that TDS responsibility can be appointed to any service based on "either the dominant-user or the most-capable-service concept" (Joint Chiefs of Staff, 1996 p. v).

Techniques, and Procedures for Movement Control, all need to be updated and enforced.

On the Air Force side, doctrine that addresses the functions of the AOC and the COMAFFOR staff needs to be updated with the concepts discussed in this report. Joint Manning Documents should be prepared to staff the positions outlined in the improved interfaces or the Deployment and Distribution Movement Organization options. The responsibilities of the AMD Assessment Cell should be written into doctrine. Any doctrine that outlines responsibilities for the A3/5, AMD, TACC, X-DDOCs, J3, or J4 will have to be revised.

Training is the key for successfully implementing the expanded STT framework. People conducting movement planning and execution processes need education in STT and multimodal movement planning and execution processes. As we have discussed, additional education and training are needed for those who will occupy key airlift planning and execution assignments that are responsible for integrating airlift into the joint end-to-end movement system. Some of these key positions will be at the USTRANSCOM DDOC, the COCOM J4 System Planning Organizations (currently X-DDOCs), the COMAFFOR A3/5/4 staff, AMDs, and the TACC. The number of positions is not large, perhaps 20 or so, but they are key to the development of the network.

For these key positions, graduate-level education is needed in multimodal movement planning and execution. Some positions should have Ph.D.-level education. Others should have master's degrees in multimodal planning and execution. AFIT could provide this education at the master's level, and civilian institutions could provide the Ph.D. education at such schools as Tennessee, Northwestern, Ohio State, and others.

Current AOC and AMD training at Hurlburt Field and the AMWC could be expanded. Taking advantage of the expertise in these training units, expanded training could include testing new tools, systems, and processes before they are fielded. STT education could be provided through Air Force continuing education at such places as the AMWC. The AMWC could expand the current AMD familiarization training to include continuing education classes on contingency network designs for those who have received education on multimodal planning and execution and STT. This class could emphasize planning in dynamic contingency environments.

Numerous opportunities exist to increase knowledge using current Air Force schools as well. The Contingency War Planners Course could increase awareness as well as attendance at the Joint Air Operations Planning course. Log 399 could provide immersion for anyone involved in J3 demand generation. Additionally, dialogue should be opened with the Army's Transportation School on opening a multimodal planning and execution course aimed at the joint end-to-end movement system development in contingency environments. A course at the Army Transportation School should have all qualified component personnel in attendance.

Finally, work should continue to expand utilization of joint exercises to train and increase operational level awareness of the development of the joint end-to-end movement system. Exercises, such as Blue Flag and Unified Engagement, could provide a venue to concentrate on movement requirements and system development implications. However, the focus of the senior leaders during the war game would need to emphasize the importance of alternative end-to-end deployment and sustainment movement options.

TO-BE Communications, Systems, and Asset Visibility Improvement Options

Many disconnects in communications exist among deployed units, APODs, the JMC (X-DDOCs), the component directors of logistics, and the AMD. Furthermore, GATES, the sustainment movement system, rides on the NIPRNET, an unclassified communications system. Many deployed sites did not have access to NIPRNET early in Operation Iraqi Freedom. And for those that did, C-130 schedules were not posted in GATES because of threat conditions. Ad hoc communications were established to furnish schedules and to track cargo that was ready to move, and so forth. A review of current and future systems and their communications architecture is needed.

Should GATES be housed on the SIPRNET so that it can meet sustainment movement needs in contingency operations? Connectivity is needed between joint common users of airlift and those who control airlift operations to meet their needs. Without an integrated end-to-end view of common information, total asset visibility and in-transit visibility are difficult to maintain—pallets could be lost. With confusion over what cargo is where, the ability to plan and replan is inhibited.

The current Air Force plan is to transfer to the Global Decision Support System (GDSS-2), which will bring together the functionality of 19 separate system and fit within the Global Combat Support System under the Global Command and Control System and the Global Information Grid architecture (see Figure 4.8).[9] The focus of our study was not to identify which information system is better than

Figure 4.8
Proposed C-130 Theater Planning Process

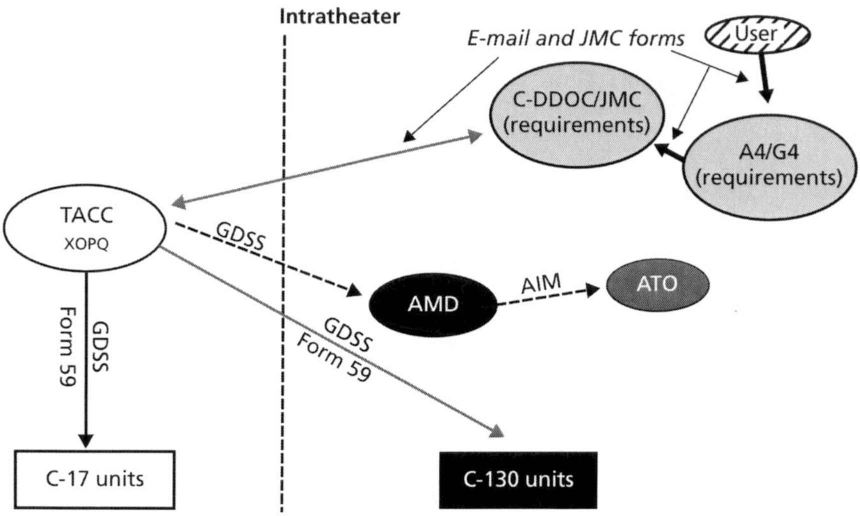

RAND *MG377-4.8*

[9] Conversation with TACC/XON personnel, November 18, 2004.

the other. We only point to the differences in the approach each service takes to identifying and monitoring multimodal movement requirements—among services as well as within individual services.

One approach for correcting these communications disconnects is to place one component in charge of standardizing the communications and systems used for communicating (airlift) movement needs (see Figure 4.9). The Air Force could accept this role, or at a minimum, play a strong role in developing communications interfaces. If it accepted, the Air Force could develop and extend systems to other military components, address the communications architecture issue, and develop uniform training packages. If the Air Force takes on this role, some adjustments in funding may be needed. There is precedence for the Air Force accepting similar responsibilities.[10] The

Figure 4.9
Communications Solutions

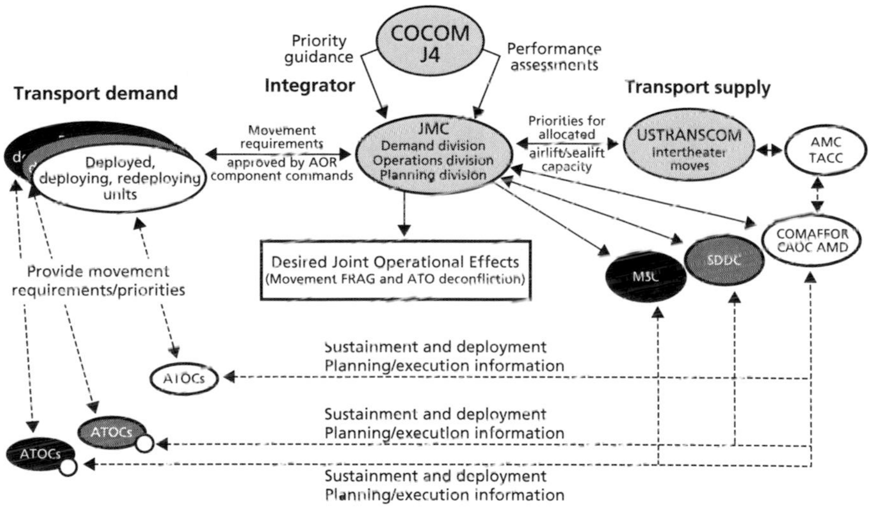

RAND *MG377-4.9*

[10] The Air Force is modifying CMOS for the Marine Corps, and they are paying to have this deployment system modified for their use.

focus of the GATES modifications needs to be on the dynamic contingency environment when FOLs are being established. Emphasis needs to be placed on getting the needed communications and systems deployed to support additional deployments and sustainment of deploying forces.

These disconnects must be addressed and corrected, whether or not the STT realignment of processes suggestion made in this report is adopted. The critical communications and information system capabilities needed to enhance TDS capability should be identified. The Joint Flow and Analysis System for Transportation sustainment planning tool needs improved data to produce more realistic sustainment requirements estimation. The ability to achieve total asset visibility and in-transit visibility are conditional on expenditures to enable RFID to work, and system upgrades are needed to enhance the transfer of data between operational units. As the data become more complete, the need to protect the data increases. SIPRNET connectivity at deployed locations could greatly enhance operations. The Air Force could be responsible for airlift communications and systems development. GATES could be the common system used for all movement requirements. Having an RFID tag read directly into GATES could solve some of the asset visibility issues.

TO-BE Improvement Summary

The enhanced J4 System Planning Organization will need strong COMAFFOR support, as well as support from each of the other components working for the COCOM. Just applying the expanded STT framework should aid the J4 in doing his or her job.

The focus of planning at the J4 level and at USTRANSCOM should be at the strategic and operational levels. This planning is focused on determining network design options, major transshipment hubs, and other major strategic and operational level decisions. This type of planning will need airlift personnel to support this level of joint planning and assessment. These airlift personnel will need extensive education in STT and multimodal movement planning and execution activities, tools, and systems.

At the component level, airlift users need to know how to access and plan for alternative airlift services. During Operation Iraqi Freedom, the components did not know how to access and use various types of airlift services until the CENTCOM LOI was developed. The LOI was effective in improving the efficiency of airlift and resulted in the reduction of many STARS routes where cargo generation was limited and could be used as a guide for contingency operations in all AORs. It could also be used to teach demand-side and supply-side processes to personnel during peacetime. The LOI identifies a set of standard processes that should be enforced in future contingencies. People with skills in surface movements—land and sea—will also need to support this enhanced joint-level planning. The Army and Navy may be able to provide some of this type of expertise.

A permanent cadre of personnel is needed in each COCOM J4 to conduct this planning on a continuing basis. This group could be supplemented by UTCs with people who have the correct education and experience. Doctrine should be revised to address these planning and assessment needs and Joint Manning Documents prepared to staff the positions. Education programs should be oriented to satisfy these needs. Some people should obtain graduate degrees in movement planning. Some courses at the AMWC could be developed to provide "SAS" equivalent "patches" for graduates. Some airlift planners should attend courses at the Army Transportation School at Fort Eustis, Va. Systems enhancements are needed for this trained cadre to schedule multimodal joint movement most effectively and efficiently.

CHAPTER FIVE

Summary and Conclusions

Closing the gap between the AS-IS theater airlift planning and execution system and the expanded STT-based framework that we have described will require significant changes to current processes, organizations, doctrine, training, and systems. Our analysis shows that improving existing processes; clearly assigning these improved supply-side, demand-side, and integrator processes to separate organizations; and enhancing doctrine, training, and systems is a preferable first step. If desired, creating a dual-hatted Deployment and Distribution Movement Organization could then be explored. Either organizational structure—assigning planning processes to existing organizations or creating a new line organization responsible for end-to-end movement planning—follows the expanded STT framework and should enhance movement effectiveness and efficiency. In terms of major policy, joint-level planning and assessment should be strengthened, and execution responsibilities should remain in the hands of the components, with improvements made in coordinating movement modes to meet dynamic requirements.

Once doctrine is in place, current processes can be revised to implement closed-loop planning and execution at the strategic, operational, and tactical planning and execution levels. Methods should be developed to trade off metrics for allocating transportation at both the global and theater levels.

Theater distribution organizational design should clearly separate supply, demand, and integrator roles. Creating J3 and J4 organizations specifically to integrate and prioritize deployment and sustainment will improve these capabilities. Placing planning and execution responsibilities for deployment and sustainment movements in one organization will facilitate development of better network options. Enhancing assessment capabilities will improve feedback on system performance and adjustments as battlefield conditions dictate. Defining reachback responsibilities and organizations will help reduce footprint and could develop new avenues for utilizing the Air National Guard (ANG). Staffing supply, demand, and integrator organizations with people trained in STT, assessment, and multimodal planning will only enhance the joint movement system.

Changes in the AS-IS TDS should be reinforced with training and exercises. Enhancing airlift planning expertise in relating movement network options and performance to operational effects can improve the quality of theater airlift planners.

In terms of communications and visibility, putting the responsibility of the communications infrastructure under one organization can help standardize procedures and eliminate command and control conflicts. Development of decision tools should improve airlift efficiency and effectiveness.

To be successful in dynamic, fluid environments, such as those associated with Operations Enduring Freedom and Iraqi Freedom, the expanded STT theater airlift planning and execution process must use the same information systems and processes. The individuals operating in the system must be trained and ready to function in the system. There needs to be standardization of systems and processes across and within theaters. Also, a process to ensure enforcement of the above requirements should be put in place and enforced.[1] This report offers solutions to all of these issues.

---

[1] Conversation with Brig Gen Mark Zamzow, CENTAF DIRMOBFOR, December 16, 2004.

APPENDIX A

Theater Distribution System Analyses Contributing Organizations

We worked with Air Force, other services, and joint stakeholders in conducting this research. Each organization openly and candidly discussed issues associated with intratheater airlift planning and execution from their vantage point. Each was interested in the framework that we used to address options for improving intratheater airlift options and ensuring that our results could be implemented.

Air Force Organizations

- Director of Mobility Forces (DIRMOBFOR)
- Air Mobility Division (AMD) personnel
- U.S. Air Forces Central Command (CENTAF)/A4
- 18th Air Force (AF)/Commander (CC)
- 21st Expeditionary Mobility Task Force (EMTF)/CC
- 15th EMTF/CC
- Tanker Airlift Control Center (TACC)/CC
- Global Mobility Concept of Operations (CONOP) Champion
- Agile Combat support (ACS) CONOP Champion
- AF/IL
- Air Warfare Center (AWC)/CC
- Combined Forces Air Component Commander (CFACC) Senior Mentors.

Joint Organizations

- U.S. Central Command (CENTCOM) Deputy CFACC
- CENTCOM JMC
- CENTCOM DDOC
- U.S. Transportation Command (USTRANSCOM)/J3/J4/J5
- U.S. European Command (EUCOM) J4 (Theater Distribution System [TDS] Conferences).

Army Organizations

- U.S. Army Central Command (ARCENT)/TDS Planners
- Arroyo

APPENDIX B

The Strategies-to-Tasks Framework and a Closed-Loop Planning and Execution Process

In this appendix, we describe the basic strategies-to-tasks (STT) framework and a closed-loop planning and execution process. The closed-loop process can be applied to critical tasks, such as allocating (and reallocating) airlift and designing (and redesigning) the movement network. Both frameworks are explained in detail in this appendix.

The RAND STT Framework

The STT framework was developed at RAND during the late 1980s[1] and has been widely applied in the Department of Defense (DoD) to aid in strategy development, campaign analysis, and modernization planning.[2] The framework has proven to be a useful approach to providing intellectual structure to ill-defined or complex problems. If used correctly, it links resources to specific military tasks that require resources, which in turn are linked hierarchically to higher-level operational and national security objectives (see Figure B.1). Working through the STT hierarchy can help identify areas where new capabilities are needed, clarify responsibilities among actors contributing

---

[1] See Kent (1989) and Thaler (1993).

[2] Internal examples are Lewis et al. (1999) and Niblack, Szayna, and Bordeaux (1996). Outside of RAND, the framework is in use by the Air Force, the Army, and elements of the Joint Staff.

Figure B.1
STT Hierarchy

RAND *MG377-B.1*

to accomplishing a task or an objective, and place into a common framework the contributions of multiple entities and organizations working to achieve some common objective.

STT Hierarchies

At the highest levels of the STT hierarchy, we consider *national goals*, which are derived from U.S. heritage and are embodied in the U.S. Constitution (see Figure B.1). These national goals do not change over time. They form the foundation from which all U.S. statements regarding national security are derived.

National security strategy is formulated by the Executive Branch. It outlines strategy for applying the national instruments of power—political, economic, military, and diplomatic—to achieve U.S. national security objectives. *National security objectives* define what must be done to preserve and protect our fundamental principles, goals, and interests with respect to threats and challenges. In contrast to national goals, national security objectives change in accordance with changes in the geopolitical environment.

National military objectives are formulated by the Secretary of Defense and the Chairman of the Joint Chiefs of Staff. The national military objectives define how the U.S. will apply military power to attain national objectives to support the national security strategy. Collectively, they define the national military strategy, which identifies (at a high level) how the United States will respond to threats to its national security.

Operational objectives describe how forces will be used to support the national military objectives. They may be regional or global and include support activities necessary to sustain military operations.

Tasks, formulated by the combatant commanders (COCOMs) and their staffs, are the specific functions that must be performed to accomplish an operational objective. Operational tasks constitute the building blocks of the application of military power. Examples of tasks that might be accomplished to help achieve the operational objective of suppressing the generation of enemy air sorties include

- crater or mine runways and taxiways,
- destroy aircraft in the open or in revetments,
- destroy key hardened support facilities, and
- destroy aircraft in hardened shelters. (Kent, 1989; Thaler, 1993.)

In this analysis, we use the STT framework to show how combat support elements, or more specifically movement capabilities, can be related to task-organized operational elements used to create desired joint operational effects by supporting the COCOM's campaign plan. In this STT construct, we outline how national goals can be broken down into national diplomatic, economic, informational, and military objectives. Regional military operational objectives can be formulated from national military objectives. Joint operational tasks can be assigned to joint task forces within the region.[3] Task-organized operational elements carry out the tasks assigned to them during different time periods, and task-organized combat support elements provide the needed support to conduct the operational mission. In

[3] The number and nature of these joint operational tasks will change over time.

this framework, movement support would be one of the task-organized combat support elements needed to create the operational effects. The movement elements could be called on to deliver troops to a given locale, to conduct operations, or to provide supplies to operational task elements needed to sustain operations in the field and, thereby, contribute to desired joint operational objectives.

The Closed-Loop Planning and Execution Process

A closed-loop assessment and feedback process[4] is a concept that has been well understood in operational planning and has been the topic of operational planning doctrine for many years (Boyd, 1987). This process can inform operational planners of how the performance of a particular combat support process affects operational capability. For example, in operations planning, it is standard procedure to conduct battle-damage assessments and, if some targets have not been destroyed or rendered unusable, to modify the air tasking order (ATO) to retarget. A schematic of the closed-loop planning and execution process is shown in Figure B.2.

The process begins, as shown on the left side of the figure, with the development of an integrated operational and combat support plan. This plan specifies the operational measures of effectiveness (MOEs) to be achieved through combat support activities—for example, F-15 weapon system availability objectives. Performance control parameters based on these operational MOEs are defined for combat support processes to create the desired operational MOE—for example, maintenance repair times, movement times. The jointly developed plan is then assessed to determine its feasibility according to availabilities of combat support resources. If the plan is infeasible, operational and/or combat support portions of the plan are identified for replanning, as shown by the closed-loop planning portion of the process on the left side of the figure.

---

[4] A *closed-loop* process takes the output and uses it as an input for the next iteration of the process.

Figure B.2
Closed-Loop Planning and Execution Process

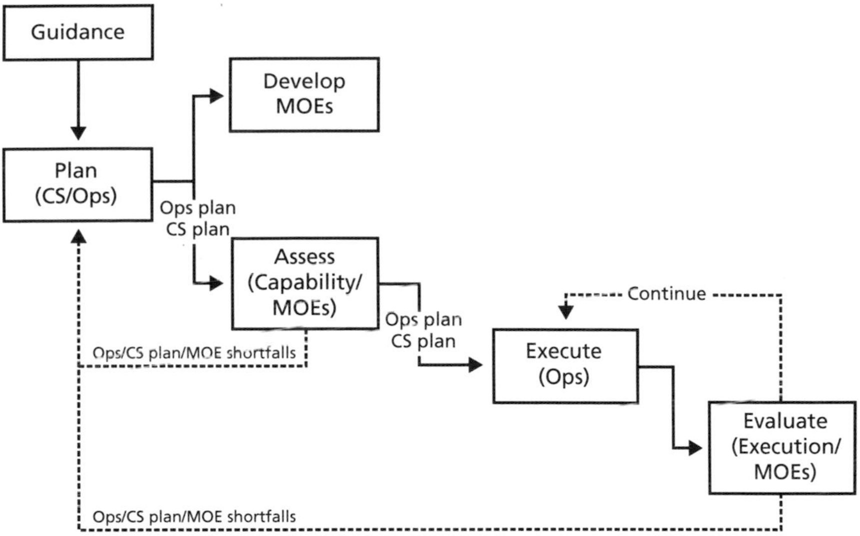

RAND *MG377-B.2*

Once a feasible plan is established, the jointly developed plan is then executed. In the execution portion of the process, actual performance of a combat support process is compared to the process control parameters identified in the planning process, as shown in the lower right of the figure. When a combat support parameter is not within the limits set in the planning process, combat support planners are notified that the process is outside accepted control parameters so that plans can be developed to get the process back within control limits.

The process centers on integrated operational/combat support planning and incorporates activities for continually monitoring and adjusting performance. A key element of planning and execution in the process template is the feedback loop, shown by the output being fed back in as input, which determines how well the system is expected to perform (during planning) or is performing (during execution) and warns of potential system failure. It is this feedback loop, which includes feedback from senior leaders, that tells the logistics and installations support planners to act when the combat support plan and infrastructure should be reconfigured to meet dynamic operational requirements, both during planning and during execution. Combat support organizations need to be flexible and adaptive so that they can make changes in execution in a timely manner (Tripp et al., 2004).

We adapt and apply this closed-loop framework to theater airlift planning and execution, focusing on tailoring force packages, allocating lift, and designing the movement network.

APPENDIX C

Closed-Loop Planning and Execution Example

Movement and Support Options

We provide an example of how the closed-loop planning and execution framework could be used to assess movement and support options. We use the Commander of Air Force Forces (COMAFFOR) A4 view of supporting F-15s deployed to southwest Asia (SWA) as an illustration—a demand-side analysis. In this example, we pose a scenario where F-15s are deployed to three forward operating locations (FOLs) for a contingency or deterrence mission. These deployed F-15s have Readiness Spares Packages (RSPs) that have been computed based on these deployed units not deploying intermediate repair capability for fixing failed avionics equipment, but rather having spares in the RSP based on sending reparable avionics components to a Centralized Intermediate Repair Facility (CIRF) and having serviceable replacements for the RSP provided from the CIRF.

Sortie generation capability is a function of many parameters including removal rates of avionics components, maintenance throughput of the repair facility—in this case a CIRF—and movement capacity and throughput capability—for example, airlift frequency between the CIRF and the deployed F-15 locations and speed of the movement pipeline for these components.

From a component viewpoint, several questions and assessments should be performed by the A4 to provide insights to the CFACC on the adequacy of the proposed or existing movement network system. These questions include the following:

- Can the proposed or existing movement system support F-15 sortie generation capability needed by the CFACC?
- If not, what are operational impacts?
- What options are available to address the shortfalls?

Figure C.1 shows the specifics of how movement performance can be related to sortie generation capability.

As indicated in this figure, flying F-15s will result in avionics component failure. These failed components will be replaced with serviceable components shown in this figure. The reparable components are replaced by serviceable components from the CIRF, or depot, which are transport time away. Shop components are needed to repair the components removed from the aircraft. These shop components are supplied from the depot. The levels of avionics components deployed to the FOLs are determined by the transport time between the CIRF and the FOLs and the repair time for components at the CIRF. The levels of shop components is dependent on the transport time between the CIRF and the depot.

Figure C.1
Closed-Loop Example (One)

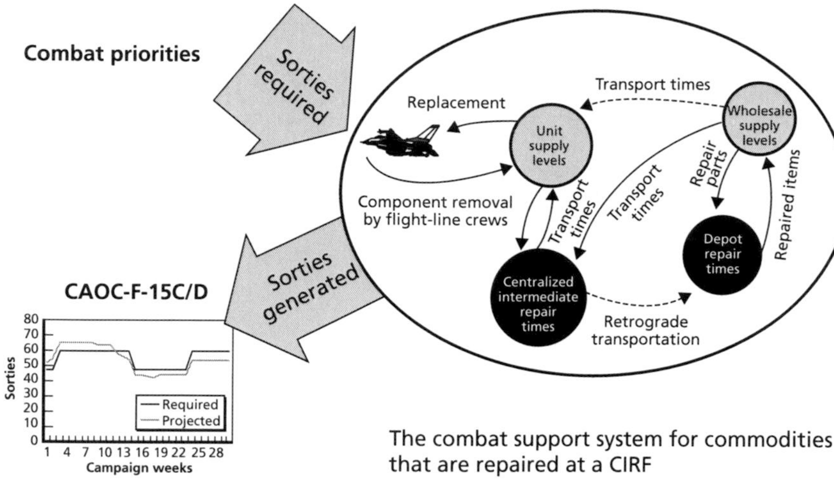

RAND *MG377-C.1*

Mathematical relationships and formulas have been developed to relate aircraft availability objectives and sortie generation capability and the support parameters discussed above. The Dyna-METRIC Microcomputer Analysis System is used by deploying units to determine levels of avionics components to take on deployments to meet specific aircraft availability objectives, given the repair concept and expected resupply (transport and processing) times.[1] In essence, this wartime spares computation system contains the combat support war plan and planned combat support performance parameters needed to meet the operational availability objectives called for by the CFACC. Specifying the planned movement performance parameter is a key part of the TO-BE movement planning and execution process.

From a CFACC perspective, the above discussion indicates how resupply movement parameters can be derived. Similar processes can be used to specify deployment times needed to create functioning FOLs.

The bottom of Figure C.2 lays out the parameters shown in Figure C.1. Data needed to track performance against each of these supply chain parameters are routinely collected, but they are not compared against performance levels needed to achieve specific operational objectives. This stems from a lack of people with the education to understand how the supply system contains wartime combat support performance parameters necessary to achieve specific operational objectives—for example, weapons system availability objectives.

In the middle of Figure C.3, we show some outtakes of data concerning transport performance consistent with data routinely collected by USTRANSCOM as part of its Strategic Distribution program. Such data can be obtained routinely from USTRANSCOM Web sites. The dotted line on both outtakes illustrates goals needed to support the F-15 weapon system availability objectives. This kind of analysis is not routinely performed by A4s.

---

[1] For more on METRIC, see Sherbrooke (1966). For more on Dyna-METRIC, see Hillestad (1982).

Figure C.2
Closed-Loop Example (Two)

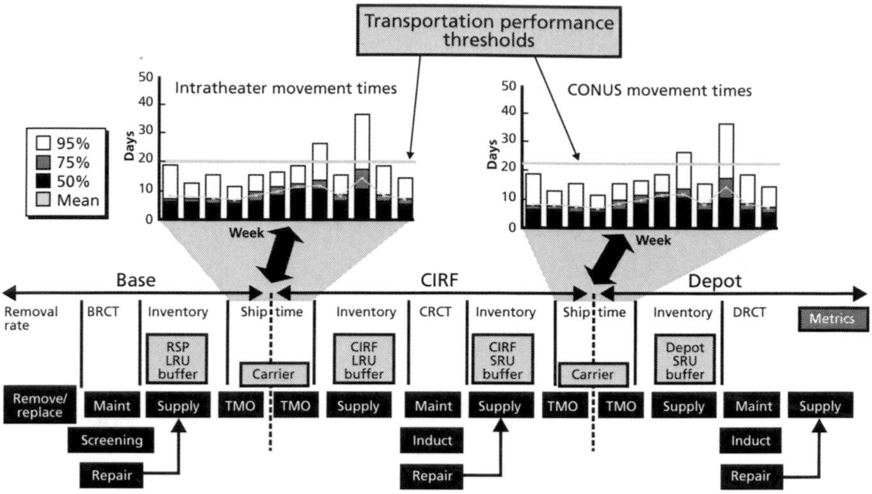

RAND *MG377-C.2*

Figure C.3 shows that when intratheater transport times breach the movement control limit, it can be expected to have an impact on F-15 sortie generation capability. Technologies to accomplish this type of assessment have been developed, but they are not routinely used.[2]

Given the signal that intratheater transport times have crossed the planned performance needed to support operational objectives, the A4 can take several actions. The first is to approach the J4 organization responsible for planning and executing TDS activities with these assessments and request support for improved transport service with the support of the CFACC. The requests could include suggestions for improvement—for example, improved frequency or alternative transport routings.

---

[2] For example, the Weapon System Management Information System.

Figure C.3
Closed-Loop Example (Three)

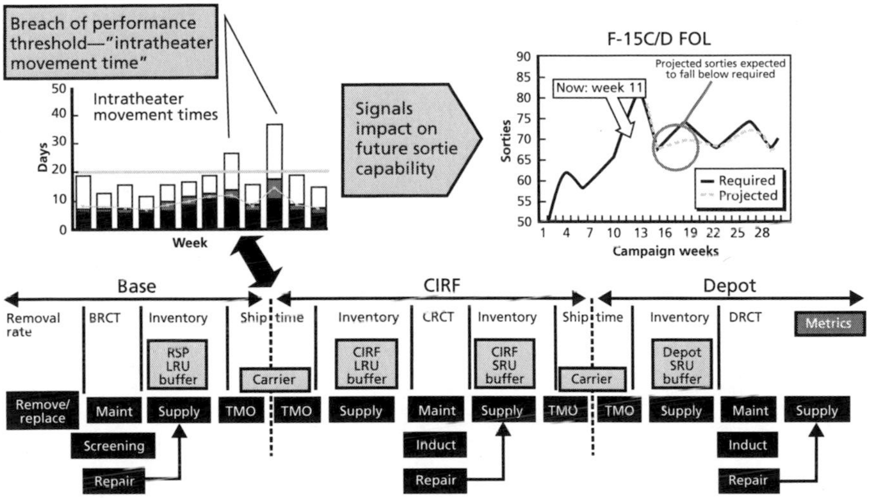

RAND *MG377-C.3*

If transport improvements can not be made, for any number of reasons, then the A4 and each deployed unit could take actions to readjust levels. If there are not enough assets to adjust deployed unit component levels and transport service cannot be improved, the operational plan may have to be adjusted.

APPENDIX D

Illustrative Example of Reachback in the Air Mobility Division

Much of airlift planning can be done through reachback. Indeed, reachback is a DoD-wide emphasis. Recently, the CENTAF AMD established reachback support with the TACC. The TACC currently supports the AMD by processing diplomatic clearances and routing C-17s assigned to move intratheater cargo and passengers in SWA. The Air Mobility Element (AME) in the AMD, however, schedules these strategic airlift assets.

Many other AMD products and services could be supplied through reachback. Reachback could be extended to C-130 and tanker scheduling. However, personnel would need to move away from three-day advance-notice, static planning to more dynamic one-day planning. Training and education would aid in making this shift.

While recent operations have shown that reachback is possible and useful, the task now is to determine the products and services that do not require intensive face-to-face interaction. Reachback is currently underutilized. Reachback can significantly reduce the AMD forward deployment footprint and offer economies of scale by providing these services with fewer people. Personnel slots gained by reducing AMD operations could be used to mitigate the need for additional airlift planners in the A3/5, J4, and USTRANSCOM.

As a note, there may be opportunities to use the Air National Guard (ANG) to support the AMD through reachback. The ANG has successfully exploited reachback to perform first level intelligence data exploitation and provided this data to the CENTAF Combined Air and Space Operations Center (CAOC).

In fact, AMC has developed a CONOP for Air and Space Operations Center (AOC)-AMD augmentation and is working with the ANG to determine the extent to which the ANG can support this mission. Other RAND work with the ANG explores AMD reachback in more detail (Tripp et al., 2004). That work developed a rule base to determine which products and services might be good candidates for reachback. We use that method in this analysis. (See Appendix E for more information about the rule-based decision tree applied in Tripp et al. [2004].)

Analysis of Reachback Options

We evaluate the costs and effectiveness of three options for implementing the improved processes in CENTCOM. Numerous variants of these options could be evaluated, however. These three options demonstrate the types of trade-offs that the Air Force faces in implementing the improved processes. We address total requirements for each option, including personnel, communications, infrastructure, deployment, and sustainment costs, and compare them with a baseline.

Option One realigns existing processes and responsibilities and separates supply, demand, and integrator responsibilities within current organizations. As shown in Table D.1, we add three personnel to develop integrated and prioritized deployment and sustainment movement requirements in the J3/5 Requirements Integration Organization. We transfer six slots from the J3/5 to the J4. These positions are associated with deployment planning at CENTCOM. These authorized slots would be combined with those working sustainment movement planning in the J4 and combined with resources allocated to the Joint Movements Center (JMC) and Deployment and Distribution Operations Center (DDOC) to form the basis of the J4 Movement Planning Organization.

Table D.1
Analysis of CENTCOM Options

	Baseline	Option 1	Option 2	Option 3
Personnel requirements				
Joint staff—total (Air Force)	1,542 (600)	1,548 [+3] (603)	1,548 (603)	1,548 (603)
J3—total (Air Force)	131 (43)	128 [–6+3] (46)	128 (46)	128 (46)
J4 (DDOC) total (Air Force)	121 (38)	127 [+6] (38)	127 (38)	127 (38)
A3/5	34	36 (+2)	36 (+2)	36 (+2)
AMD (forward)	93	94 (+1)	60	60
AMD (rear at TACC)	0	0	34 (+1–3)	31 (+1–7)
Total—joint (Air Force) net	NA (NA)	+3 (+3) +6	+3 (+0) +3	+3 (–4) –1
Infrastructure	NA	NA	$92,000	$84,000
Communications enhancements	NA	NA	Adequate	Adequate
Reduced annual deploy cost	NA	NA	$300,000	$300,000
Reduced annual sustain cost	NA	NA	$400,000	$300,000
One-time reduction (tents)	NA	NA	$400,000	$300,000
Potential intratheater effectiveness improvements (10%)	NA	UTC for 3 C-130s	UTC for 6 C-130s	UTC for 6 C-130s

We estimate that no net gain in personnel is needed to accomplish the functions of the J4 Movement Planning Organization that we described. Rather, the personnel are focused on strategic- and operational-level planning and shifted away from problem-solving associated with tactical planning and execution. Option One also increases the number of authorizations in the A3/5 by two to enhance strategic- and operational-level airlift planning and to support the J4 System Planning Organization.[1] This option also increases the authorization to conduct airlift assessments by one over the current assignment of one person. This person is identified with future operations in the AMD staffing document.

Option Two has most of the airlift scheduling processes being performed from the TACC in a dedicated cell for the COMAFFOR. This option leaves Airlift Control Team ATO production support

[1] A3/5 personnel data provided by 9th Air Force/USCENTAF Manpower office.

and liaison officer support forward (five personnel).[2] The AME support is moved to the TACC. This option, as well as the third, keeps the DIRMOBFOR and other key staff, including tanker scheduling,[3] forward where face-to-face interaction is needed with the CFACC, but has the routine airlift scheduling performed from the rear. In this option, we assume no improvements are made in processes. We also assume that no overhead economies are associated with providing reachback support from the TACC although there could be modest differences in personnel authorization associated with combining support functions already provided by the TACC that include a slight reduction in overhead. However, those saving are not illustrated here.

Option Three changes the reachback organization from dedicated COMAFFOR cells to one organized by planning and execution functions. This option also considers savings possible through the application of shareware that make it possible to conduct scheduling of multiple areas of responsibility (AORs) simultaneously. We estimate efficiencies associated with this modest software enhancement coupled with improved collaborative tools, improved scheduling processes, and a better understanding of the entire requirement could drive efficiencies of about 10 percent for a total reduction of roughly three people (34 × 0.1). This would leave about 31 people at the TACC to run CENTAF operations on three shifts and to cover for vacations and other absences.[4]

Using the reachback decision tree developed for the ANG Transformation Project (Tripp et al., 2004) and Air Force Flight

---

[2] AMD personnel data is from an air base unit manning document.

[3] In this example, tanker scheduling is kept forward. Tanker scheduling failed the "Is this product stable?" question in the decision tree for reachback (in Appendix E) (Discussions with 15th EMTF personnel, October 2004).

[4] The deployment requirements manning document for the CENTAF AMD shows a total of 93 authorizations. We realize that the number of authorizations for the AMD has been as high as 140 and is currently running about 80 authorizations. We use the AEF 3 and 4 authorizations to illustrate how reachback and other process improvements may impact personnel requirements.

Manual AN/USQ-163-1, Falconer,[5] we identified 34 positions whose functions could be performed via reachback from the TACC. We also eliminated nine positions: one AMD supervisor, three in the AME, three communications, and two supply positions whose functions can be provided by existing TACC personnel. We left 60 positions forward with the CAOC. These positions include: the DIRMOBFOR and most of his staff of eleven positions; eight support positions including force protection and intelligence; eight positions in the Aerial Refueling Control Team whose responsibilities include the planning and replanning of tanker missions; five positions in the Aeromedical Evacuation Control Team (AECT) whose responsibilities include central source for expertise for aeromedical evacuation missions; and 17 positions associated with the Air Mobility Control Team (AMCT). The AMCT is the centralized source for command, control, and communications for air mobility forces during execution. Of the 60 slots that we left forward at the CAOC, we suggest there may be more potential to perform more services from the TACC saving the Air Force further.

Table D.1 also shows a rough cost estimate for implementing the improved interfaces options for CENTCOM in terms of total and Air Force personnel authorizations. Option One requires additional Air Force authorizations, while Options Two and Three require fewer authorizations than the baseline.

We calculated additional infrastructure costs needed at the TACC to support intratheater airlift planning and execution. We allowed 15 square feet per person and used the planning cost associated with new construction costs ($2,715 per person) cited in the *Historical Air Force Construction Cost Handbook* (AFCESA, 2004).[6] Option Two would require roughly $92,000 to house 34 additional people (34 × $2,700). Option Three would require $84,000 (31 ×

---

[5] Definitions of Air Mobility Division and teams taken from Air and Space Operations Center, AN/USQ-163-1, Block 10, Version 1-0, November 26, 2002.

[6] According to AFCESA (2004, p. 9), the average cost of new headquarters facility was $181 per square foot. We multiply the $181 by 15 to determine the cost per person of $2,715.

$2,700). The cost to refurbish the TACC may be less than the cost for new construction.

According to the personnel at the 18th Air Force TACC,[7] sufficient communication infrastructure is available to require no additional communication funding increases.

We estimate the reduction in annual deployment costs for both Options Two and Three to be $300,000 per year for airlift costs.[8] We estimate $400,000 reduction in annual sustainment costs for Option Two and an approximate $300,000 reduction for Option Three.[9] We estimate one-time cost savings associated with not purchasing tents of $400,000 for Option Two and a $300,000 savings for Option Three.[10]

The last row of Table D.1 shows an estimate of effectiveness improvement that could be brought about primarily through improving the integrated estimates and better prioritization of air movements resulting from separating demand and supply relationships on the joint staff. We show a modest 5 percent improvement, or a reduction in three C-130s that would need to deploy. The academic literature shows that another 10 percent improvement in effectiveness could be expected if scientific algorithms were used to support routing and scheduling decisions (Armacost, Barnhart, and Ware, 2002; Cohn and Barnhart, 2003; Weigel and Cao, 1999; Mingozzi, Baldacci, and Ball, 2000). In today's ad hoc implementation of the AMD, it may be difficult to achieve these effectiveness increases. If the TACC, a permanent organization, had this responsibility, it could be argued that these additional enhancements to effectiveness resulting from improved routing and scheduling may materialize. We

---

[7] Interview with TACC/XON personnel, 2004.

[8] The annual deployment cost reduction would be: $56 dollars per person per hour for airlift times 20 hours (flight time) times 34 (the number of people not deploying) times 2 (deployment and redeployment) times 4 (the number of 90 day rotations in a year) or $\$56 \times 20 \times 34 \times 2 \times 4 = \$300,000$ per year in transport savings.

[9] Sustainment estimate of $34 \times \$30$ per day $\times 365 = \$400,000$.

[10] $5.6 million is the cost for one Harvest Falcon 550 Housekeeping set. So, the cost of not having to house 34 people would be $\$5.6 \times (34 \div 550)$ = approximately $400,000.

show an additional 5 percent improvement in effectiveness under the two reachback options to account for this likelihood.

Table D.2 extends the bottom line analysis for CENTCOM and shows the impacts for implementing the improved interface options in the three major AORs—that is, EUCOM, U.S. Pacific Command (PACOM), and CENTCOM combined. In Table D.2, we use the UTCs to support the initial response package for PACOM and EUCOM. These UTCs are designed to support 300 sorties and have 66 authorizations each. In this example, we have the same functions being performed through reachback at the TACC as in the CENTCOM illustration. This would have the following functions being performed through reachback from the TACC: AME functions (11 authorizations in each UTC); most Airlift Control Team functions (8 of 10 authorizations in each UTC—two are left forward for ATO production support). We eliminate the 11 AME positions when moved to the TACC. We add one assessment cell person for each area. This would have nine authorizations moved to the TACC for PACOM and EUCOM contingency support; 47 people would deploy forward to support the CFACC, and 11 authorizations would be eliminated.

Table D.2
Extended Analysis of CENTCOM Options

	Baseline	Option 1	Option 2	Option 3
Personnel requirements				
AMD forward	225	228	148	148
AMD rear at TACC	0	0	49	44
Total—joint (Air Force) net	0 (0)	+9 (+9) +18	+9 (–16) –7	+9 (–22) –13
Infrastructure	NA	NA	$130,000	$120,000
Communications enhancements	NA	NA	Adequate	Adequate
Reduced annual deploy cost	NA	NA	$400,000	$400,000
Reduced annual sustain cost	NA	NA	$500,000	$500,000
One-time reduction (tents)	NA	NA	$500,000	$500,000
Potential intratheater effectiveness improvements (10%)	NA	UTC for 6 C-130s	UTC for 12 C-130s	UTC for 12 C-130s

For Option Two this would mean that (31 + 9 + 9) 49 would not deploy. The TACC would have a total of 49 people supporting reachback options over three shifts.

Option Three includes a 10 percent improvement in efficiency as assumed in the CENTCOM analysis. This option would have 44 people at the TACC.

Options Two and Three have the same adds and subtracts to the PACOM and EUCOM Joint staffs and COMAFFOR A3/5 staffs as the CENTCOM analysis.

The remainder of the table is filled out using the same assumptions used in the CENTCOM analysis.

APPENDIX E

Reachback Decision Tree

For reachback analysis, RAND developed a tool for nominating potential reachback candidates. This tool is a decision tree (see Figure E.1). The decision tree can be applied to any task. It is a series of questions to which the answer is yes or no. The answer to a question routes the user down the tree until reaching the end of a branch. The end of the branch will either offer the task as a potential candidate or eliminate the task for reachback. The decision tree itself is an Access database that tracks a user's answers and provides a way to capture comments and/or assumptions about the questions and/or answers (Tripp et al., 2004).

Figure E.1
Reachback Decision Tree

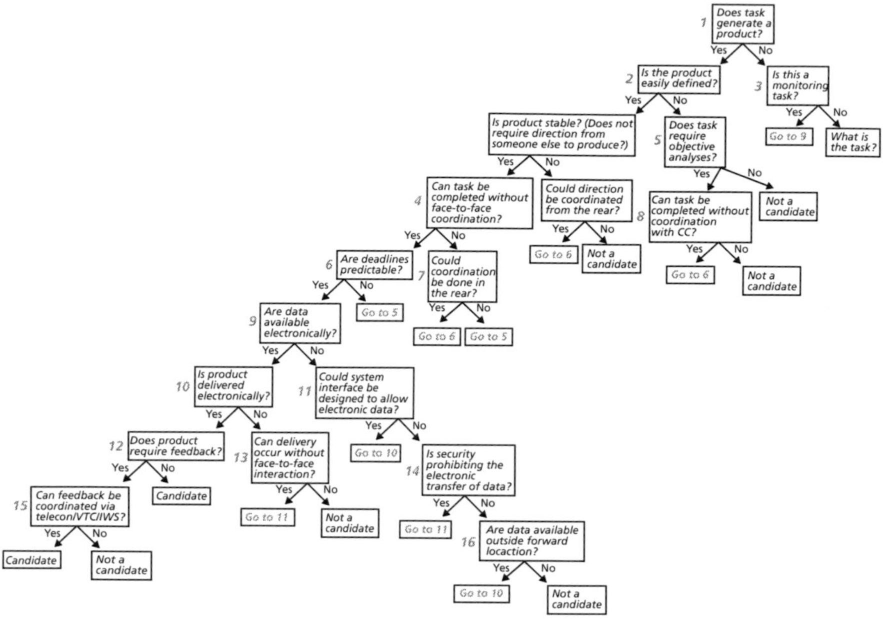

RAND *MG377-E.1*

APPENDIX F

Evolved CENTCOM Intratheater Airlift Planning Process

CENTCOM's intratheater airlift planning process has evolved since the beginning of operations in that area of responsibility (AOR). When this study began, CENTCOM processes followed doctrine as prescribed. However, problems occurred with cargo backlog and development of standard air routes. Senior leaders took steps that moved planning processes more in line with the options listed in Chapter Four of this report. The evolved intratheater airlift planning process is outlined in Figure F.1.

Figure F.1
CENTCOM AOR Evolved Intratheater Airlift Planning Process

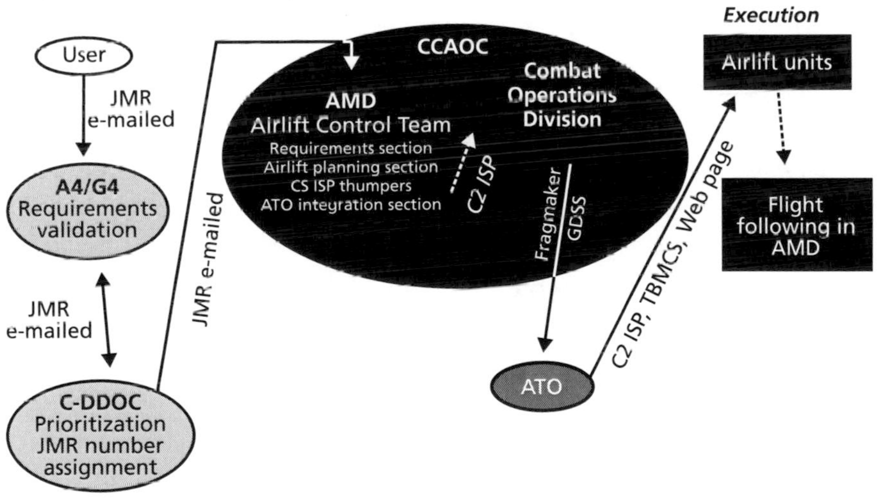

RAND *MG377-F.1*

The user could be a forward deployed unit or part of a Combined Joint Task Force (CJTF) element. The user would submit a request to either their respective component A4/G4 or to the CJTF-4 for validation of the movement request by air. The request is made using the Joint Movement Request (JMR) form (see Figure F.2). The requestor enters all required information including point of origin, destination, number of pallets, short tons to be moved, names of passengers, and any information on hazardous cargo. The request is then attached to an e-mail and sent to the respective A/G/CJTF-4 for validation as a priority air movement. Once validated by the respective A/G/CJTF-4, the JMR is again attached to an e-mail and forwarded to C-DDOC for assignment of an airlift priority and a JMR routing number. The CENTCOM combatant commander (COCOM) has developed 17 airlift priorities for airlift. They range from mail and other morale items through critical supplies and ammunition. The C-DDOC then attaches the JMR to an e-mail and forwards the request to the requirements section of the Airlift Control Team (ALCT) who is a part of the Air Mobility Division (AMD) in the CENTCOM Combined Air and Space Operations Center (CAOC). The requirements section then hand-transfers the data from the JMR into the Cargo Planning Fragmentation Order (FRAG), an Excel spreadsheet developed for airlift planning.

Once in the Cargo Planning FRAG, the information is combined with other inputs to create a suggested FRAG for the upcoming days. The data are then sent electronically to the Plans Division of the ALCT, which quality-checks the requirements, assigns requirements to specific flying organizations, and assigns call signs. The information is then forwarded to the C2 ISP personnel who transcribe the data from the Cargo Planning FRAG into C2 ISP (yet another system). Once in C2 ISP, the data flows on the *low* or unclassified side to the air tasking order (ATO) integration section of the ALCT. ATO integration ensures that the data flow to both the high (classified) and low side of C2 ISP as well as flowing from high C2 ISP into the theater battle management control system where the data are integrated with all other ATOs.

Once validated by the Combat Plans Division of the CAOC, the data become the final, published ATO. Once the final ATO is published, the flying units can view the tasking through various systems. The data continue to reside in C2 ISP and is available throughout the world. The Combat Plans Division also publishes the data on a CENTCOM-specific Web page that can be viewed in the theater.

During execution, the Air Mobility Control Team (AMCT) provides flight following and works with the requirements section of the ALCT to attempt to ensure that efficient and effective changes are made when required. However, a lack of total asset visibility combined with poor communication capability and a fluid and dynamic environment oftentimes preclude the best solution.[1] For example, suppose a C-130 aircraft were scheduled to pick up passengers at a specific APOD and an opportune airlift C-17 happened to deliver cargo and transport the passengers out of the APOD prior to the C-130's arrival. Even if the AMD were aware of the movement, communications capability in the theater combined with diplomatic clearances and time required to attain permission to operate out of another airfield may preclude the AMD from even attempting to make any changes to the assigned C-130 schedule.

A new information system is being tested in the theater—the Intratheater Transportation Request System—as part of the AMC Combined Air Mobility Planning System. While it will streamline the requesting process, replacing the JMR and easing the ability for the users to track their requests through the approval process, this system is an AMC program, not a joint program. Issues remain about how well this will be accepted outside of the CENTCOM AOR. Additionally, the Intratheater Transportation Request System does not address all of the manual entering of information that is required once an approved request reaches the AMD.

---

[1] Conversations with Air Mobility Division personnel, December 2004.

Figure F.2
CENTCOM Joint Movement Request Form

UNCLASSIFIED FOR TRAINING USE ONLY

CENTCOM AOR / JMC MOVEMENT REQUEST FORM JMC#

SECTION 1

Unit Point of Contact:		DATE:	
POC SIPR Address:		DSN Phone:	
POC NIPR Address:		Comm Phone:	
UNIT & REASON:			
ROUND TRIP or ONE-WAY		Continuation Sheets Used ?	

COMPONENET COMMAND'S WILL PERFORM QUALITY CONTROL CHECK AND ASSIGN UNIT TRACKING NUMBER PRIOR TO SUBMITTAL

UNIT TRACKING NUMBER		QC CERTIFIER	

SECTION 2

FILL IN ALL INFORMATION AVAILABLE AT TIME OF REQUEST.

1. READY TO LOAD DATE (RLD) = ALD - 1:		6. POE = ORIGIN:			12. # PAX	
2. AVAILABLE TO LOAD DATE (ALD):		7. POD = DEST:			13. BULK S/T	
3. EARLIEST ARRIVAL DATE (EAD) = ALD+1:		8. ULN/PID:			14. OVERSIZE S/T	
4. LATEST ARRIVAL DATE (LAD = RDD):		9. UIC:			15. OUTSIZE S/T	
5. REQUIRED DELIVERY DATE (RDD):		10. # PALLETS		11. # ROLLING STOCK	16. TOTAL S/T	0

SECTION 3

PALLETS / CONTAINERS/LOOSE CARGO (If additional rows are needed use the Pallet/Container List TAB provided below) **LIST ALL PALLETS INDIVIDUALLY**

DESCRIPTION OF CARGO:	LGTH:	WDTH:	HGT:	WGT:	ULN/TCN:	HAZMAT Y/N

ROLLING STOCK (If additional rows are needed use the Rolling Stock TAB provided below) **LIST ALL ROLLING STOCK INDIVIDUALLY**

UNIT ID	VEHICLE DESCRIPTION	TCN:	LGTH:	WDTH:	HGT:	WGT:	HAZMAT Y/N

HAZARDOUS CARGO DATA(If additional Space is needed please use the bottom of the Pallet - Container List TAB prov ided below)

It is the Shipper's/Requestor's Responsibility to Identify all HAZMAT Information IAW Applicable DoD Directives

PROPER SHIPPING NAME	UN/ID #	PCS	GWT (LBS)	NEW(KG) / QUANITY	CLASS / DIV	ULN / PLT ID

PASSENGERS (If additional rows are needed use the Passenger List TAB provided below)

RANK	FULL NAME	RANK	FULL NAME

SECTION 4

ADDITIONAL INFORMATION

JMC USE ONLY

MODE OF TRAVEL		A = Air T = Truck S = Ship X = Rail	JMC Priority Number	
Aircraft / Vehicle / Vessel ID	DATE	ROUTE		

JMC COORDINATION

ACTION	DATE	TIME	PRINTED NAME OF VALIDATOR
RECEIVED BY:			
JOPES Newsgroup:			
USER CONTACTED:			
Completed/Cancelled			

JMCF02 (26 Feb 04) UNCLASSIFIED FOR TRAINING USE ONLY PREVIOUS VERSIONS OBSOLETE

RAND *MG377-F.2*

Bibliography

Air Force Civil Engineer Support Agency (AFCESA), *Historical Air Force Construction Cost Handbook*, Tyndall AFB, Fla., 2004.

Amouzegar, Mahyar, Robert S. Tripp, Ron McGarvey, Edward W. Chan, and C. Robert Roll, Jr., "Supporting Air and Space Expeditionary Forces: Analysis of Combat Support Basing Options," Santa Monica, Calif.: RAND Corporation, MG-261-AF, 2004.

Armacost, A., C. Barnhart, and K. Ware, "Composite Variable Formulations for Express Shipment Service Network Design," *Transportation Science*, Vol. 36, No. 1, February 2002.

Boyd, John R., "A Discourse on Winning and Losing," Maxwell AFB, Ala.: Air University Library, Document No. M-U4397, unpublished collection of briefing slides, August 1987.

Cohn, A., and C. Barnhart, "Improving Crew Scheduling by Incorporating Key Maintenance Routing Decisions," *Operations Research*, Vol. 51, No. 3, 2003.

Hillestad, Richard J., *Dyna-METRIC: Dynamic Multi-Echelon Technique for Recoverable Item Control*, Santa Monica, Calif.: RAND Corporation, R-2785-AF, 1982.

Joint Chiefs of Staff, Office of the Chairman, *Joint Tactics, Techniques, and Procedures for Movement Control*, Washington, D.C.: Joint Publication 4-01.3, 1996.

Kent, Glenn, *A Framework for Defense Planning*, Santa Monica, Calif.: RAND Corporation, R-3721-AF/OSD, 1989.

Lewis, Leslie, James A. Coggin, and C. Robert Roll, *The United States Special Operations Command Resource Management Process: An Applica-*

tion of the Strategy-to-Tasks Framework, Santa Monica, Calif.: RAND Corporation, MR-445-A/SOCOM, 1994.

Lewis, Leslie, Bruce Pirnie, William A. Williams, and John Y. Schrader, *Defining a Common Planning Framework for the Air Force*, Santa Monica, Calif.: RAND Corporation, MR-1006-AF, 1999.

Lynch, Kristin F., John G. Drew, Robert S. Tripp, and C. Robert Roll, Jr., *Supporting Air and Space Expeditionary Forces: Lessons from Operation Iraqi Freedom*, Santa Monica, Calif.: RAND Corporation, MG-193-AF, 2005.

Mingozzi, A., R. Baldacci, and M. Ball, "The Rollon-Rolloff Vehicle Problem," *Transportation Science*, Vol. 34, No. 3, August 2000.

Myers, Richard B., Chairman, Joint Chiefs of Staff, *National Military Strategy of the United States*, Washington, D.C.: The Pentagon, 2004.

National Security Strategy of the United States, Washington, D.C.: The White House, September 2002.

Niblack, Preston, Thomas S. Szayna, and John Bourdeaux, *Increasing the Availability and Effectiveness of Non-U.S. Forces for Peace Operations*, Santa Monica, Calif.: RAND Corporation, MR-701-OSD, 1996.

Robbins, Marc L., and Eric Peltz, "Sustainment of Army Forces in Operation Iraqi Freedom: End-to-End Distribution," Santa Monica, Calif.: RAND Corporation, MG-343-A, forthcoming.

Schrader, John Y., Leslie Lewis, William Schwabe, C. Robert Roll, and Ralph Suarez, *USFK Strategy-to-Task Resource Management: A Framework for Resource Decisionmaking*, Santa Monica, Calif.: RAND Corporation, MR-654-USFK, 1996.

Sherbrooke, Craig C., *METRIC: A Multi-Echelon Technique for Recoverable Item Control*, Santa Monica, Calif.: RAND Corporation, RM-5078-PR, 1966.

Snyder, Don, and Patrick Mills, *A Methodology for Determining Air Force Deployment Requirements*, Santa Monica, Calif.: RAND Corporation, MG-176-AF, 2004.

Thaler, David E., *Strategies to Tasks: A Framework for Linking Means and Ends*, Santa Monica, Calif.: RAND Corporation, MR-300-AF, 1993.

Tripp, Robert S., Lionel Galway, Timothy L. Ramey, and Mahyar Amouzegar, *Supporting Expeditionary Aerospace Forces: A Concept for*

Evolving to the Agile Combat Support/Mobility System of the Future, Santa Monica, Calif.: RAND Corporation, MR-1179-AF, 2000.

Tripp, Robert S., Kristin F. Lynch, John G. Drew, and Edward W. Chan, *Supporting Air and Space Expeditionary Forces: Lessons from Operation Enduring Freedom*, Santa Monica, Calif.: RAND Corporation, MR-1819-AF, 2004.

Tripp, Robert S., Kristin F. Lynch, Ron McGarvey, Raymond A. Pyles, Don Snyder, William Williams, and James M. Masters, "Strategic Analysis of Air National Guard Combat Support and Reachback Functions," Santa Monica, Calif.: RAND Corporation, MG-375-AF, forthcoming.

U.S. Air Force, Office of the Secretary of the Air Force, *Gulf War Air Power Survey,* Washington, D.C., 1993.

Walden, Joseph I., *The Forklifts Have Nothing to Do: Lessons in Supply Chain Leadership,* Lincoln, Neb.: iUniverse, Inc., 2003.

Weigel, D., and B. Cao, "Applying GIS and OR Techniques to Solve Sears Technician-Dispatching and Home-Delivery Problems," *Interfaces*, January–February 1999.